CHICAGO PUBLIC LIBRARY
HAROLD WASHINGTON LIBRARY CENTER

W9-ABC-933

REF
SF
97
.N3

National Research
Council.

Nutrients and
toxic substances
in water for...

Cop l.

DATE DUE

REF
SF
97
.N3

FORM 125 M

SCIENCE DIVISION

The Chicago Public Library

Received_____ JUN 1 7 1977 _____

NUTRIENTS AND TOXIC SUBSTANCES IN WATER FOR LIVESTOCK AND POULTRY

National Research Council · Subcommittee on Nutrient and Toxic Elements in Water

A Report of the
SUBCOMMITTEE ON NUTRIENT AND
TOXIC ELEMENTS IN WATER

Committee on Animal Nutrition
Board on Agriculture and Renewable
    Resources
Commission on Natural Resources
National Research Council

National Academy of Sciences
WASHINGTON, D.C. 1974

NOTICE: The project that is the subject of this report was approved by the Governing Board of the National Research Council, acting in behalf of the National Academy of Sciences. Such approval reflects the Board's judgment that the project is of national importance and appropriate with respect to both the purposes and resources of the National Research Council.

The members of the committee selected to undertake this project and prepare this report were chosen for recognized scholarly competence and with due consideration for the balance of disciplines appropriate to the project. Responsibility for the detailed aspects of this report rests with that committee.

Each report issuing from a study committee of the National Research Council is reviewed by an independent group of qualified individuals according to procedures established and monitored by the Report Review Committee of the National Academy of Sciences. Distribution of the report is approved, by the President of the Academy, upon satisfactory completion of the review process.

This study was supported by the U.S. Department of Agriculture.

Library of Congress Catalog Card No. 74-2836
International Standard Book Number 0-309-02312-2

Available from
Printing and Publishing Office, National Academy of Sciences
2101 Constitution Avenue, N.W., Washington, D.C. 20418

Printed in the United States of America

REF
SF
97
.N3
cop.1

JUN 17 1977

PREFACE

The Subcommittee on Nutrient and Toxic Elements in Water was requested to clarify, as much as possible, the value of water as a source of nutrients for livestock and poultry. Only limited experimental studies on this subject have been made. Nutrient substances that may at times be present in water at toxic levels should be defined in regard to concentration and conditions in which they are harmful to various species. Excessively saline waters, whether naturally derived or resulting from prior use, need evaluation. Such substances as nitrates and sulfates may be beneficial at low concentrations but are harmful at high levels. Many elements and substances that occur in water have no nutrient properties; at times these attain chronically or acutely toxic concentrations. Problems of water procurement and purification, as well as pollution in general, were considered to be too extensive to be covered in this study, which was prepared under the direction of the Committee on Animal Nutrition.

Committee on Animal Nutrition

    T. J. Cunha, *Chairman*

    J. P. Bowland

    C. W. Deyoe

    W. H. Hale

    J. E. Halver

    E. C. Naber

    R. R. Oltjen

    L. H. Schultz

    R. G. Warner

Subcommittee on Nutrient and Toxic Elements in Water

    R. L. Shirley, *Chairman*

    C. H. Hill

    J. T. Maletic

    O. E. Olson

    W. H. Pfander

CONTENTS

INTRODUCTION

Extensive analyses of surface water and groundwater by geological, agricultural, and health agencies have demonstrated the common presence at varying concentrations of all the known essential dietary mineral elements for livestock and poultry. This is largely due to the so-called universal solvent property of water for polar substances. However, the literature contains only limited documentation that these nutrients in water are available in amounts that would be a dietary aid in modern livestock and poultry production.

The importance of water for cattle and other animals was reviewed by Leitch and Thomson (1944) and by Sykes (1955). Winchester and Morris (1956) used data in the literature to work out tables of water consumption for cattle of different ages. They adjusted their values according to the animals' body size, feed intake, lactation, butterfat, and ambient temperature. When they plotted water intake per unit of dry matter ingested against ambient temperature, water intake changed very little from –12 to 4 °C but accelerated from 4 to 38 °C. For instance, a 450 kg steer would drink 28, 49, and 66 liters daily at 4, 21, and 32 °C, respectively.

Harbin et al. (1958) observed a high correlation between water intake and temperature, but at constant temperatures relative humidity had no effect. When Balch et al. (1953) reduced water intake to 60 percent of that drunk in a control period, five of six cows refused part of their hay diet and all lost weight. Utley *et al.* (1970) reported that a restriction of water to 60 percent of free choice among Angus steers resulted in a decrease in digestibility of crude fat but an apparent increase in digestibility of dry matter, protein, and nitrogen-free extract. Maximum work energy expenditure in dogs weighing 8–12 kg and aged 2.5 years was reported by Young *et al.* (1960) to be a function of water intake within the range of 0–1 liter/day; intakes of 1.2–1.6 liters/day were required to obtain full work potential. Nicholson *et al.* (1960) added alfalfa ash to a low quality roughage ration for calves and observed that it increased daily water intake.

Water was shown by Medway and Kare (1958) to alleviate salt toxicity. Pigs and chickens could be killed experimentally by drenching them with NaCl solutions and depriving them of drinking water, while those having access to unlimited water supply suffered no ill effects. Wells that penetrate salt deposits, lakes in arid areas, and irrigation water runoff are sources of saline water that may exceed the salt tolerances of livestock.

Alsmeyer *et al.* (1955) fed four pigs per treatment a corn–soybean-meal ration from 32 to 91 kg weight with water *ad libitum* from pond, well, and distilled sources. The animals allowed well water required 7.7 kg less feed per 45.5 kg gain than those consuming distilled water. With pond water, 2.7 kg less feed was required.

Increased physical concentrations of both livestock and poultry, combined with heightened emphasis on nutrition and health, have raised the number of inquiries about water quality. The safety factor of water has become a greater concern as the chances of pollution by the public, industry, and agriculture itself increase.

This publication summarizes information on the effects of known nutrient and toxic substances that occur in water consumed by domestic animals. It also discusses data concerning water requirements, the percentages of the recommended intake of various required elements and substances that may be provided in normal water intake per day, and the concentration of substances that may occur at toxic levels for various species.

DISSOLVED AND SUSPENDED CONSTITUENTS IN WATER SUPPLIES

The Geochemical Cycle

The world's supply of water remains constant at about 273 liters for each square centimeter of the earth's surface (Goldschmidt, 1933). This amount is made up of 268.45 liters of seawater, 0.1 liter of fresh water, 4.5 liters of continental ice, and 0.003 liter of water vapor. This fixed supply moves along pathways through the hydrologic cycle illustrated in Figure 1. Movement within the cycle produces progressive changes in the water. Mist from waves may contribute most of the minerals returned to the atmosphere from oceans. Such processes as oxidation, reduction, cation and anion exchange, precipitation, dissolution, acid–base and coordinative interactions, microbial transformations, evaporation, transpiration, and erosion continually alter the dissolved substances in water supplies. Water is thus never pure; dissolved constituents are always present as derived from the natural environment and from the activities of man.

The weathering of rocks and concurrent migration of water through their interstices, as well as movement of water through soil, are major factors in the alteration of water composition. The magnitude and kinds of changes occurring are controlled by rock types, kinds of soils, and energy available to drive the reactions. Water percolating through soil and rocks containing an abundance of soluble minerals, such as saline

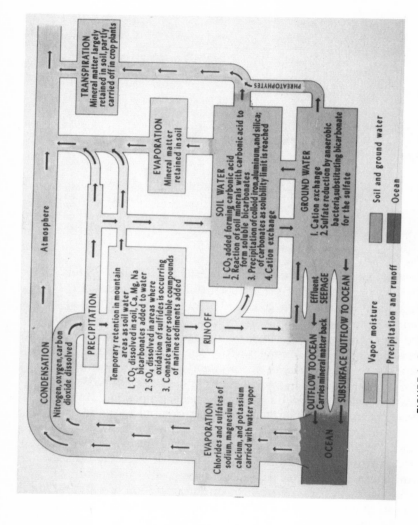

FIGURE 1 Geochemical cycle of surface and groundwaters (Davis et al., 1959).

4

soils and marine shales, will have a higher dissolved-solids load than water percolating through clean sands and gravels containing mostly sparingly soluble minerals. Back and Hanshaw (1965) identify the primary controls on the dissolved solids content of groundwater as being the original chemical character of the water as it enters the zone of saturation; the distribution, solubility, and exchange capacity and exchange selectivity of the minerals in the deposits; the porosity and permeability of the rocks; and the flow path of the water.

The operation of the geochemical cycle for a particular locale is described by Davis *et al.* (1959). The cycle, as applied to the San Joaquin Valley, California, includes:

1. movement of airmasses over the ocean toward land carrying small but significant quantities of dissolved solids, particularly sodium, chloride, and iodide ions;

2. condensation of water-carrying nitrogen, oxygen, and carbon dioxide from the atmosphere to the ground in rain or snow;

3. entry of water into the soil, accompanied by further enrichment with carbon dioxide;

4. dissolution of minerals with attendant release of cations and anions;

5. solubilization of compounds, such as sodium chloride and gypsum, causing a direct addition to the salt load;

6. oxidation of sulfides in organic sediments to sulfates, providing additional amounts of this ion;

7. exchange of cations within the soil solution and groundwater for those in soils and rocks, precipitation of compounds as solubility products are exceeded, and bacterial reduction of sulfate in solution; and

8. return of water to the atmosphere by evaporation or transpiration, leaving the chemical products behind; or return of water to the ocean as streamflow or groundwater discharge carrying with it the various ion species, suspended colloidal materials, and perhaps some sediment.

Water in open lakes and reservoirs primarily reflects the compositions of influent water and is subject to dynamic cycles of diurnal, seasonal, and annual changes. These changes are generated by the action of organisms, reactions at the mud–water interface, sediment inflow, and variations in heat energy.

Many other factors, in addition to soil and rock types, can cause important differences in water composition. Thus, industrial solid waste and gaseous discharges may add significant quantities of minerals and organic complexes. Feedlots may be the source of ammonia (Hutchinson and Viets, 1969). Seacoast regions receive larger amounts of sodium and chloride ions from rainfall than areas farther inland. The dissolved-solids load arising from urban sources usually adds about 300 mg/liter. In addition to dissolved minerals, water may carry dissolved gases, oily substances, phenols, detergents, pathogens, and other undesirable substances that can adversely affect animal health and the quality of their products.

Constituents in Water Supplies

Water sources of primary use to livestock and poultry include small streams, rivers, lakes, impoundments, and groundwaters. These inland waters contain most of the elements occurring in the soils, sediments, and rocks of the earth. Table 1 (Davis and DeWiest, 1966) classifies 58 mineral constituents into major, secondary, minor, and trace categories. Ten elements make up approximately 99 percent of the dissolved mineral compounds (Garrels and Christ, 1965). These elements are hydrogen, sodium, potassium, magnesium, calcium, silicon, chlorine, oxygen, sulfur, and carbon. They can be considered to occur in solution as ions, molecules, or radicals and may be designated as H^+, Na^+, K^+, Ca^{2+}, Mg^{2+}, CO_3^{2-}, HCO_3^-, Cl^-, SO_4^{2-}, H_4SiO_4, and $H_3SiO_4^-$. Simi-

TABLE 1   Dissolved Solids in Potable Water—A Tentative Classification of Abundance[a]

Major Constituents (generally less than 1,000 mg/liter)				
Bicarbonate	Chloride	Silica	Sodium	Sulfate
Calcium	Magnesium			
Secondary Constituents (generally less than 10.0 mg/liter)				
Boron	Fluoride	Nitrate	Potassium	Strontium
Carbonate	Iron	Phosphate		
Minor Constituents (generally less than 0.1 mg/liter)				
Aluminum	Cadmium[b]	Germanium[b]	Manganese	Titanium[b]
Antimony[b]	Chromium[b]	Iodide	Molybdenum	Uranium
Arsenic	Cobalt	Lead	Nickel	Vanadium
Barium	Copper	Lithium	Rubidium[b]	Zinc
Bromide				
Trace Constituents (generally less than 0.001 mg/liter)				
Beryllium	Gold	Radium	Silver	Tungsten[b]
Bismuth	Indium	Ruthenium[b]	Thallium[b]	Ytterbium
Cerium[b]	Lanthanum	Scandium[b]	Thorium[b]	Yttrium[b]
Cesium	Niobium[b]	Selenium	Tin	Zirconium[b]
Gallium	Platinum			

[a]Position of phosphate and selenium changed to conform to findings 1957–1969 STORET Program Summary (Systems for Technical Data, 1971).
[b]Those elements that occupy an uncertain position in the list.

larly the remaining elements will occur as individual ions, pairs of ions, or complexes of several ions. Some examples are Fe as Fe^{2+}, Fe^{3+}, $FeOH^+$; N as NO_2^-, NO_3^-, NH_4^+; As as $HAsO_4^{2-}$, $H_2AsO_4^-$, H_3AsO_4; Cr as $Cr_2O_7^{2-}$, CrO_4^{2-}, $CrOH^{2+}$; and P as PO_4^{3-}, HPO_4^{2-}, $H_2PO_4^-$, H_3PO_4, $H_2P_2O_7^{2-}$, $HP_2O_7^{3-}$, $P_2O_7^{4-}$. Identification of the ionic form of the element is important in assessing nutritional or toxic effects of a few elements. Known instances of this phenomenon are indicated in the sections on nutrients and toxic elements in water.

The measurement of the concentration of all constituents dissolved in water is referred to as "total dissolved solids" (TDS). The term "salinity" as applied to fresh waters is often used synonymously with "total dissolved solids" as a statement of the total ionic concentration.

A universal classification of water according to salinity levels has not been developed. Various disciplines have developed classifications to meet their specific needs. Two general classifications were developed by Davis and DeWiest (1966) and by Robinove *et al.* (1958) for the United States Geological Survey (USGS). The former is:

Description	Concentration of TDS (mg/liter)
Fresh water	0–1,000
Brackish water	1,000–10,000
Salty water	10,000–100,000
Brine	> 100,000

The latter is:

Description	Concentration of TDS (mg/liter)
Slightly saline	1,000–3,000
Moderately saline	3,000–10,000
Very saline	10,000–35,000
Brine	> 35,000

Total dissolved constituent concentrations range from about 25 mg/liter in areas of insoluble rock types with high rainfall to more than 300,000 mg/liter where saturated sodium chloride solutions occur. TDS provide a useful index to the suitability of a water supply for livestock use.

Water Properties

The constituents in a given water impart certain characteristic properties that assume special relevance when related to particular water uses. Such properties include hydrogen-ion activity (pH), alkalinity, acidity, electrical conductivity, hardness, color, turbidity, biological oxygen demand, chemical oxygen demand, taste, odor, sodium absorption ratio, radio-activity, and density. Temperature may also be regarded as a property since it influences physical, chemical, and biological reactions, thereby affecting the behavior and use of the water.

The pH, defined as the negative logarithm of hydrogen-ion activity, is a measure of the effective concentration of hydrogen ions in water. At very low concentrations, activity and concentration are practically the same. Below pH 7 water is acidic (excess of hydrogen ions), and above that value it is alkaline (excess of hydroxyl ions).

Natural waters are known to have a pH value generally in the range of 6 to 9. Values above 9 do occur in springs under unusual conditions. Values below 6 also occur under circumstances usually involving sulfur oxidation reactions, as in some mine drainage. The pH influences taste, corrosivity, efficiency of chlorination, and other water-treatment processes. As a factor in itself the hydrogen-ion activity does not directly affect animal nutrition, but it does serve to screen water that may present problems, particularly when the value lies outside the 6–9 range.

Water "hardness" has been understood to indicate the tendency of water to precipitate soap or to form scale on heated surfaces. Hardness is generally expressed as the sum of calcium and magnesium reported in equivalent amounts of calcium carbonate. Other cations, such as strontium, iron, aluminum, zinc, and manganese, also contribute to hardness and, when present in unusual amounts, should be determined and included in the computation. Most surface waters have hardness values of less than 1,000 mg/liter while groundwaters generally have less than 2,000 mg/liter. In arid regions these values may be higher. The hardness of domestic water, according to Durfor and Becker (1964), may be classified as:

Hardness Range (mg/liter)	Description
0–60	Soft
61–120	Moderately hard
121–180	Hard
> 180	Very hard

Hardness per se is not a problem in livestock drinking water. Generally, the concentrations of the individual ions that may be nutrients or toxicants are important.

Water Used by Livestock

As estimated in *Water Quality Criteria* (1968), in the United States about 1.7 billion gallons of water per day is withdrawn for livestock uses. Almost 60 percent of this amount is estimated to be withdrawn via wells from groundwater sources. The remainder is derived from streams, lakes, springs, and impoundments such as stockponds.

THE WATERS OF STREAMS

The composition of stream water is characterized by spatial and temporal variabliity. Controlling factors include the climatic setting, particularly the rainfall composition of water as it enters the soil mantle; the kinds of soils, unconsolidated sediments, and rock types through which it percolates; and the influences of man's activities. Water in the upper reaches of streams and along tributaries reflects local geochemistry. As flow proceeds downstream the local chemical differences are integrated. As a result, the principal dissolved solids in downstream portions of larger river systems tend to resemble one another (Livingstone, 1963). The concentration of dissolved solids broadly correlates with rainfall. Figure 2, as prepared from Rainwater's (1962) compilation of prevalent or modal dissolved solids concentrations, shows that the lowest concentrations occur in areas of highest precipitation. Most of the streams along the East Coast and the southeastern portion of the Gulf of Mexico, in mountainous areas and the uplands of Lake Superior, and along the Pacific Northwest coast have concentrations of less than 100 mg/liter.

Adjacent to and farther inland from these regions—extending to and

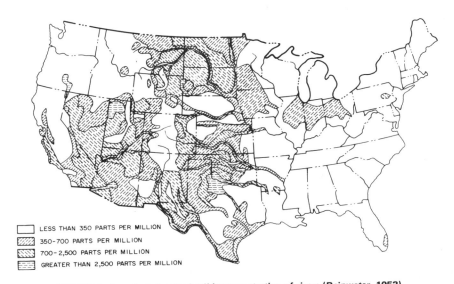

LESS THAN 350 PARTS PER MILLION
350-700 PARTS PER MILLION
700-2,500 PARTS PER MILLION
GREATER THAN 2,500 PARTS PER MILLION

FIGURE 2 Prevalent dissolved solids concentration of rivers (Rainwater, 1952).

including the western Gulf of Mexico coastal plain, the Ozark plateau, eastern portions of the central lowlands, the Nebraska sandhills, and the Columbia plateau—concentrations are generally in a range of about 100–300 mg/liter. Streams draining portions of the northern and southern Great Plains; the southern portion of the Osage plains, basin, and range; and the Southwest exhibit a wide diversity of concentration that may range from 300 to about 6,000 mg/liter. Throughout much of this area the rainfall is low and the geology diverse. Soluble salts have not been as strongly leached; hence larger amounts of soluble material are dissolved as the water percolates through the weathered zone.

The variations in the principal kinds of constituents in streamflow have been categorized and displayed in Figure 3, also developed by Rainwater (1962). He grouped the waters into four chemical types, each being characterized by their most prevalent constituents. These types are Ca–Mg/CO_3–HCO_3, Ca–Mg/SO_4–Cl, Na–K/CO_3–HCO_3, and Na–K/SO_4–Cl. Eighty-seven percent of the water in the conterminous United States is dominated by calcium and magnesium, while only 13 percent is of the sodium–potassium type. Calcium and magnesium usually occur with carbonates and bicarbonates and, to a lesser extent, with sulfate and chloride. Sodium and potassium are more likely to be combined with sulfates and chlorides than with carbonates and bicarbonates.

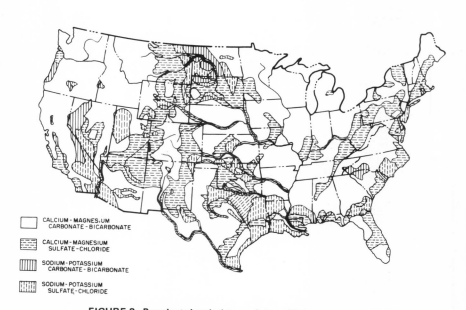

CALCIUM - MAGNESIUM
CARBONATE - BICARBONATE

CALCIUM-MAGNESIUM
SULFATE-CHLORIDE

SODIUM-POTASSIUM
CARBONATE - BICARBONATE

SODIUM- POTASSIUM
SULFATE-CHLORIDE

FIGURE 3 Prevalent chemical type of rivers (Rainwater, 1962).

Examples of analyses of selected streams representing the four chemical types are presented in Table 2. These data more fully characterize the dissolved constituents carried in water supplies. The importance of local variations in water composition, particularly in small streams and tributaries, should not be overlooked in assessing water quality for livestock use. For example, in the northern Atlantic Coast drainage net, major streams will have dissolved solids concentrations of less than 350 ppm, while a smaller stream, such as Shamokin Creek, at low discharge, may contain a concentration of about 1,000 ppm. Similar comparisons may be made for other drainage nets. When great diversity exists in soils and rocks within a basin, particularly in arid and semiarid zones, a wider variation in the water quality of small streams can be expected.

Temporal, as well as spatial, variability in stream quality may also be great. Generally, the relationship between stream discharge and total dissolved solids is inverse. The components of streamflow are surface runoff and groundwater flow. The groundwater component tends to remain stable. Thus, during periods of low streamflow concentrations are maximum because most of the discharge is from groundwater. Such base flows have had a long residence time in the soils and rocks, favor dissolution, and are therefore more concentrated than the surface runoff, which has a much shorter residence time or which may be largely overland flow.

During storms the infiltration rate of soil decreases as the soil moisture content increases. Consequently, some storm intensities will exceed, for given periods, the infiltration rate and issue large volumes of overland runoff to the stream. This runoff will thus have a low dissolved solids concentration. Such storm runoff may, however, carry a heavy sediment load; when it does, cation exchange between the water and sediment minerals may further alter the composition of the water.

Variations in water quality with respect to discharge are most pronounced in the arid and semiarid zones, where maximum and minimum concentrations of dissolved solids at a specific site may differ by a factor of 5–10 or even larger (Rainwater, 1962). In the wetter areas, where water is subjected to leaching over a longer geologic timespan, the variation is much smaller. For example, Livingstone (1963) reports that the range between minimum and maximum total dissolved solids concentrations for the Mayo River, North Carolina, is only 37–57 mg/liter.

LAKES AND PONDS

Lakes and ponds form an important source of water for livestock. According to Reid (1961), most open lakes have a 100–200 ppm concen-

TABLE 2 Examples of Stream Waters of Various Chemical Types

River	Location	Cations (mg/liter)				Anions (mg/liter)				TDS (mg/liter)
		Ca	Mg	Na	K	HCO$_3$	SO$_4$	Cl	NO$_3$	
Calcium–magnesium, carbonate–bicarbonate										
Flathead	Columbia Falls, Mont.	23	6.0	1.1	1.3	95	4.5	0.8	0.4	138
Withlacooche	Holder, Fla.	44	3.8	5.0	0.3	118	23	10	0.7	213
Calcium–magnesium, sulfate–chloride										
Pecos	Alamagordo Dam, N. Mex.	366	54	60	4.7	95	1,020	90	1.0	1,710
Eagle	Gypsum, Colo.	55	11	26	3.5	107	101	38	1.3	351
Sodium–potassium, carbonate–bicarbonate										
White	Kadoka, S. Dak.	18	1.7	108	5.7	249	76	5.7	1.8	509
Amargosa	Beatty, Nev.	2.0	0.2	423	17.1	639	257	109	0.5	1,520
Sodium–potassium, sulfate–chloride										
Arkansas	Dardanelle, Ark.	44	9.6	95	3.5	104	53	159	3.0	481
Cimarron	Perkins, Okla.	200	39	1,670	–	136	474	2,600	–	5,190

tration of dissolved solids. Closed lakes, on the other hand, may have concentrations in excess of 100,000 ppm, due primarily to the concentrating effect of evaporation and the continued inflow of solutes. Hutchinson (1957) compiled data for 132 closed lakes from his own and other studies and found the following distribution:

Concentration (mg/liter)	Percent of Total
10–100	0
100–1,000	25.8
1,000–10,000	25.8
10,000–100,000	31.8
> 100,000	16.6

As the concentration of such lakes increases, the dominant ions normally shift from calcium to sodium and from bicarbonates to sulfates and chlorides. This results from the precipitation of calcium carbonate and calcium sulfate as their solubility products are exceeded.

Compared to streams, open lakes generally tend to be more chemically stable and exhibit less variations in composition and concentration (Livingstone, 1963). As the water is detained, mixing occurs, thus minimizing variations within the lake and in the outflow. Major influences on the concentration of dissolved solids arise from the composition of the inflow, volume of inflow and outflow, volume of precipitation and evaporation, the energy budget, biological activity, and lake geometry. The chemistry of the inflow is the most important of these; therefore, the dissolved constituents can be expected to be similar to that of the streams in the drainage basin. Local aberrations, of course, may arise. Thus, near oceans comparatively dilute lakes may contain more Na^+ and Cl^- than Ca^{2+} and HCO_3^-. Solute salts may be brought into solution from the reservoir basin; saline springs may contribute salts within the lake itself.

Most open lakes, particularly in temperate zones, undergo dynamic diurnal, seasonal, and annual changes in the concentration of many of the dissolved ions, primarily such elements as oxygen, carbon, hydrogen, nitrogen, phosphorus, silicon, sulfur, iron, calcium, magnesium, sodium, and potassium. Many of these participate in complex cyclical transformations, such as the nitrogen, phosphorus, iron, sulfur, and silica cycles. As a result, both temporal and spatial distributions of the chemical constituents occur within the lake. These changes are detailed in classical limnological texts such as Hutchinson's (1957).

Table 3 presents some examples of the chemical composition of lake waters. While sampling was not carried out uniformly, the data suffice

14

TABLE 3 Chemical Composition of Selected Lakes in the United States by Regions

Lake	Location	Cations (mg/liter)				Anions (mg/liter)				TDS (mg/liter)
		Ca	Mg	Na	K	HCO₃	SO₄	Cl	NO₃	
Southern Atlantic coast										
Okeechobee[a]	Clewiston, Fla.	41	9	22	1	136	28	29	1.2	277[b]
Great Lakes										
Superior[a]	Several places	14	3	—	3.4	50	4	1	0.5	82[b]
Erie[a]	Huron, Ohio	39	8	8	1	121	28	17	1.2	227[b]
Upper Mississippi										
Traverse[f]	Wheaton, Minn.	129	99	100	20	184	750	21	8.5	1,330
Pokegama[f]	Grand Rapids, Minn.	31	13	3	1	164	7	0	1.0	163
Lower Mississippi										
Corney[i]	Summerfield, La.	38	5	93	4	19	1	220	0.2	452
Concordia[i]	Ferriday, La.	27	9	6	2	135	0	7	0.0	137
Missouri										
Byron[h]	Huron, S. Dak.	52	94	343	40	269	922	131	0.2	1,810
Cottonwood[h]	Agar, S. Dak.	51	38	202	39	370	379	59	0.4	976
Devils Lake										
Devils[a]	Devils Lake, N. Dak.	60	306	1,680	176	565	3,460	787	2.7	7,050
Free Peoples[a]	Devils Lake, N. Dak.	80	92	2,810	104	1,768[c]	3,600	800	3.0	9,300
Stink[a]	Devils Lake, N. Dak.	41	590	6,370	185	507[c]	13,000	1,670	2.1	22,400
Rio Grande										
Balmorhea[a]	Jeff Davis Co, Tex.	38	0	642[k]		158	555	560	0.2	1,970
Ft. Stockton[a]	Pecos Co., Tex.	186	3	544[k]		153	968	378	0.2	2,260
La Sal Vieja[a]	Willacy Co., Tex.	308	71	8,250[k]		156	995	13,090	0.3	22,900

Great Salt Lake										
Mud[d]	Montpelier, Idaho	21	10	13	5.9	103[c]	16	13	–	150[b]
Salt Lake[j]	Utah	324	8,400	89,400	5,040	332	20,200	151,000	69	278,000
Inland basins										
Salton Sea[a]	Imperial Co., Calif.	505	581	6,249	112	232	4,139	9,033	1.2	20,900
Little Borax[a]	Lake Co., Calif.	8	24	3,390	731	8,166[c]	<10	905	–	13,600
Borax[a]	Lake Co., Calif.	Nil	30	6,199	332	6,668[c]	21	5,945	–	>19,400
Mono[a]	Mono Co., Calif.	11	32	21,400	1,120	26,430	7,530	15,100	2.4	71,900
Pons at Bad Water[a]	Death Valley, Calif.	1,230	148	14,100	594	187[c]	4,960	21,400	–	42,700
Winnemucca[a]	Nevada	20	18	1,321	70	571	135	1,725	1.6	3,890
Pyramid[a]	Stutcliffe, Nev.	10	113	1,630	134	1,390[c]	264	1,960	2.2	5,510
Eagle[a]	Susanville, Calif.	4	49	220	59	870[c]	1.6	22	0.4	1,250
Lower Alkali[a]	Eagleville, Calif.	6	0.9	1,370	11	1,200[c]	307	1,160	3.5	4,150
Middle Alkali[a]	Cedarville, Calif.	17	8.9	3,180	7.5	2,040[c]	576	3,330	0.3	9,240
Albert[a]	Valley Falls, Oreg.	7	2.9	6.5	2.1	53	1.2	1.2	–	108
Hot[g]	Washington	640	22,838	7,337	891	6,296	103,680	1,668	–	>143,000
Tahoe[g]	Bijou, Calif.	9	2.3	6.4	1.9	51	2	3	0.1	63
Columbia										
Park[a]	Coulee City, Wash.	30	18	34[k]		212	29	12	1.8	380
Lenore[a]	Soap Lake, Wash.	3	20	5,360[k]		9,110[c]	2,180	1,360	3.2	18,000
Lowell[e]	Idaho	31	5	35	2	159	25	15	–	–

[a]Livingstone, 1963.
[b]Analysis represents mean of several samples; all others are single samples.
[c]Includes CO_3.
[d]Jensen et al., 1951.
[e]Lewis, 1959.
[f]USGS Water Supply-Paper 1643, 1959a.
[g]USGS Water Supply-Paper 1645, 1959b.
[h]USGS Water Supply-Paper 1743, 1960.
[i]USGS Water Supply-Paper 1884, 1961a.
[j]USGS Water Supply-Paper 1885, 1961b.
[k]Values are sum of sodium and potassium concentrations.

to demonstrate that important chemical differences exist among lakes in various climatic and geochemical settings. As in the case of streams, much variability occurs in the lakes of the inland basins and the Southwest, where considerable local leaching of soluble salts can occur in the lake drainage basin. Otherwise, open lakes tend to have a composition similar to that of streams in the area. Hutchinson's compilations also show that about one out of every two closed lakes has a water quality that is undesirable for livestock use.

All lakes age and are subject to eventual extinction. This process is natural but can be hastened by man's activities, such as discharge of municipal, industrial, and agricultural wastes. The complex of processes causing aging is referred to collectively as eutrophication. Initially, a lake is characterized by low concentrations of plant nutrients and minimal biological productivity; it is an oligotrophic lake. As the plant nutrients increase, biological productivity increases and the lake becomes mesotrophic. Continuation of this sequence leads to a highly productive, or eutrophic, lake. As extinction approaches, the remaining water body becomes a pond, marsh, or swamp. Through this succession of stages, the dissolved species change. It is not unusual for lakes in close proximity to be in different stages of eutrophication. For example, many small lakes on the glacial terrain of eastern South Dakota can be found in various stages of aging.

Because of their capacity to produce large algal blooms, eutrophic lakes are the least desirable for stock watering. While harvested green algae may serve as a food supply for livestock, certain blue-green algae are toxic and can cause death when ingested. According to Greeson (1970), the green algae become dominant during late spring but decrease with the arrival of the blue-green algae, which thrive during the summer, when maximum temperature and light conditions prevail. Other properties of water in an eutrophic lake, such as taste and odor, also detract from its usefulness.

Important sources of surface water for livestock are ponds and small reservoirs. Such sources have not been extensively studied with respect to their dissolved constituents. The generalizations given above for the major factors influencing water quality in lakes are also applicable to small ponds.

Sloan (1970), in a study of prairie potholes of glacial origin in the north–central United States and south–central Canada, found that differences in climate, geology, topography, groundwater hydrology, and land use tend to cause wide variations in the dissolved constituents of these small water bodies. The potholes ranged in size from a fraction of an acre to several square miles and were usually less than 1 m (3 ft) deep and seldom exceeded depths of 1.8 m (5 ft). Salinity and depth were found to fluctuate rapidly within an individual pothole both seasonally and annually in response to inflow and outflow. Retention of water in the potholes ranged from a few days following spring thaw to a more or less permanent state. In these regions the potholes are an important source of water supply for livestock and are among the best wetland habitats for waterflow breeding.

Table 4 presents the concentration of the principal dissolved constituents of some of the typical waters as found by Sloan (1970). The dissolved solids concentration was found to range from very fresh waters to brines that were several times more concentrated than seawater. The fresh waters are dominated by Ca^{2+} and HCO_3^-, brackish waters by Mg^{2+} and SO_4^{2-}, and saline waters by Na^- and SO_4^{2-}.

Small ponds on the southern Great Plains were investigated by Toetz (1967). These ponds are widely used for livestock watering. In the 29 ponds studied, the mean annual concentration of dissolved solids ranged from 87 to 359 mg/liter and averaged 174 mg/liter. Seasonal variations were also observed. Spring rainfall caused a decrease in concentration, followed by a sharp increase in July. Subsequent evaporation apparently caused continued increases in the dissolved constituents. Although the principal constituents in the water were not measured, the reported mean annual pH and alkalinity values for individual ponds suggested important chemical differences.

Playa lakes are found in essentially barren and flat regions that are also generally dry and undrained. They may contain water of shallow depth for short periods of time and many are salty. Studies of 41 playa lakes typical of those that eventually collect most of the runoff from the high plains of Texas were made by Wells *et al.* (1970). Complete analyses for the principal dissolved constituents were not made; estimates, however, could be made for most of the lakes. These showed that the dissolved solids concentrations were mostly below 300 mg/liter during much of the year. Wide exceptions were recorded; for example,

TABLE 4 Principal Dissolved Constituents in Prairie Potholes Developed on Stagnation Moraines in North Dakota

Pothole Water	Cations (mg/liter)				Anions (mg/liter)			Calculated Dissolved Solids (mg/liter)
	Ca	Mg	Na	K	HCO_3	SO_4	Cl	
Fresh	34	12	2.8	24	144	31	6.9	254
Brackish	250	534	470	85	473	3,260	71	5,450
Saline	729	2,700	4,540	510	355	17,910	2,590	29,200

the concentration in one lake was about 4,100 mg/liter in July 1970 and only 83 mg/liter in May 1969. Other lakes showed about a tenfold difference between minimum and maximum concentrations. Important seasonal shifts in the predominant cations and anions also occurred. Sodium usually exceeded calcium concentration and bicarbonates and sulfates exceeded chloride levels.

In other studies of playa waters on the southern high plains, Lotspeich *et al.* (1969) found the water to be satisfactory for livestock use. Moreover, they concluded that cultural practices in the contributing areas of the playas did not appreciably influence water composition. A playa with a watershed containing 95 percent native grass had water that was chemically similar to water in playas whose watersheds were nearly all cultivated.

GROUND WATERS

Groundwater is the major source of water for livestock in the United States. Whenever water that infiltrates the soil mantle exceeds the evapotranspirational demand, deep percolation occurs. A groundwater body forms when the rate of this percolation is sufficient to saturate the available porosity above a barrier layer. The pore space in which the groundwater resides has three principal regimes. Davis and DeWiest (1966) identify these as the intergranular, fracture, and tubular, or solution, regimes. In the first, the water is stored in the primary pore space (as in consolidated or poorly consolidated sediments); the second, in cracks or fissures (as in igneous, metamorphic, or firmly consolidated sedimentary rocks); and the third, in large openings or solution channels (as in volcanic flow and carbonate rocks). The salinity of groundwater is related to these porosity regimes and to the chemical composition of the rocks and sediments.

Temporal variations in groundwater quality are usually small and take place gradually over a period of years. Spatial variability, however, is great and is, of course, related to the soil and rock types in a given local area. In unconsolidated and poorly consolidated sediments, lateral movement of water usually predominates over horizontal movement. As a result of vertical barriers impeding flow, water quality may differ vertically. Thus, the water available for use can in some areas depend on the depth of well placement. The yield of water from fractured rock is generally very low, and the quality in such groundwater bodies is variable and strongly controlled by the input quality and the rock type. In the volcanic flow and carbonate rocks, the channels typically transmit large volumes of water and wells are often highly productive. The

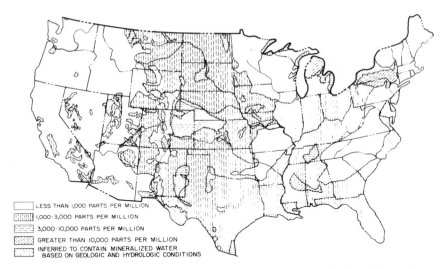

LESS THAN 1,000 PARTS PER MILLION
1,000-3,000 PARTS PER MILLION
3,000-10,000 PARTS PER MILLION
GREATER THAN 10,000 PARTS PER MILLION
INFERRED TO CONTAIN MINERALIZED WATER BASED ON GEOLOGIC AND HYDROLOGIC CONDITIONS

FIGURE 4   Total dissolved solids concentration of shallow groundwater (Adapted from Feth et al., 1965).

dissolved solids content of the water in carbonate rocks is usually high due to dissolution phenomena. Conversely, the low solubility of volcanic rocks limits dissolution; therefore, the dissolved solids content is generally low.

Areal differences in the quality of shallow groundwater are presented in Figure 4 (Feth et al., 1965). The map shows that most of the shallow groundwaters in the United States have dissolved solids concentrations of less than 3,000 mg/liter and should be satisfactory within the limitations discussed in the section on salinity of water as related to livestock production (pp. 39–50). Table 5, compiled from studies by White et al. (1963), relates typical compositions and concentrations of the major dissolved constituents to rock types and demonstrates the influence of many rock formations on water composition. These influences, however, may be modified by such actions as the intrusion of seawater as coastal aquifers are depleted, the accretion of soluble salts from the leaching of saline overburden, and the recycling of the aquifer supply for irrigation and through recharge with polluted waters.

Minor and Trace Constituents

In natural waters the principal source of the minor and trace constituents (as defined in Table 1) is the release of soluble products during the

TABLE 5 Chemical Composition of Subsurface Waters from Selected Rock Formations in the United States

Rock Type	Location	SiO$_2$ (mg/liter)	Cations (mg/liter)				Anions (mg/liter)				TDS (mg/liter)
			Ca	Mg	Na	K	HCO$_3$	SO$_4$	Cl	NO$_3$	
Granite, rhyolite, and similar rock types											
Silicic volcanics	Grandview, Idaho	37	3.6	0.8	3.9	2.3	21	2.6	1.4	1.9	75
Rhyolite	Burns, Oreg.	62	14	5.8	20	5.2	112	72	4.0	3.1	234
Granite	West Warwick, R.I.	20	6.5	2.6	5.9	0.8	38	0.9	5.0	1.5	82
Gabbro, basalt, and ultramafic rock types											
Gabbro	Waterloo, Md.	39	5.1	2.3	6.2	3.2	37	9.2	1.0	0.3	112
Basalt	Moses Lake, Wash.	53	29	19	12	3.5	177	15	6.9	9.7	328
Olivine basalt tuff-breccia	Buell Park, Ariz.	31	20	42	19[a]		279	22	7.0	2.5	423
Sandstone, arkose, and graywacke											
Catahoula sandstone	Collins, Miss.	25	2.4	0.5	2.6	2.0	18	1.4	2.5	0.0	55
Dawson arkose	Monument, Colo.	35	9.6	1.9	5.1[a]		38	7.4	1.8	1.5	101
Rensselaer graywacke	Sand Lake, N.Y.	12	74	20	34	1.2	381	26	2.7	1.0	553
Siltstone, clay, and shale											
Tuffaceous clay	Sheaville, Oreg.	43	29	11	31	8.0	180	33	6.0	0.7	342
Clay	Georgetown, S.C.	12	1.6	0.7	210[a]		479[b]	1.5	28	0.0	735
Ohio shale	Park Lake, Ky.	22	15	7.5	36	3.5	3.2[b]	128	21	0.5	239
Pierre shale	Langdon, N. Dak.	26	416	143	362	14	104[b]	2,170	38	0.1	3,300
Limestone and dolomite											
Miocene limestone	Gainesville, Fla.	10.0	15.0	6.7	3.2	0.6	74	2.6	3.4	1.8	118
Gasconada dolomite	Alley, Mo.	5.4	30.0	18.0	4.6[a]		164	1.4	5.0	0.9	229

20

Miscellaneous sedimentary rocks

Rock type	Location										
Biwabik iron formation	Grand Rapids, Minn.	14	54	19	7.5	5.8	271	6	0.5	1.2	380
Big Fork chert	Hot Springs, Ark.	26	26	1.9	7.4	2.8	68	34	2.2	0.1	171
Gypsum (castle form)	Red Bluff, N. Mex.	29	636	43	17[a]		143	1,570	24	18	2,480
Quartzite and marble											
Mutual quartzite	Kamas, Utah	3.6	2.6	0.4	1.0	1.8	8.0	3.4	0.8	1.2	23
Quartzite	Cliffs Shaft Mine, Mich.	7.6	32	16	8.5	3.1	144	39	8.0	2.7	262
Sylacauga marble	Sylacauga, Ala.	9.9	39	10	2.7	0.3	162	2.4	3.8	5.8	236
Slate, schist, gneiss, and greenstone											
Siamo slate	Morris Mine, Mich.	8.2	101	5.6	13	10	212	128	5.2	0.0	489
Mica schist	Wilkesboro, N.C.	26	10	1.6	5.5	1.0	45	3.0	2.5	1.4	99
Gneiss	Nipton, Calif.	30	90	69	72	2.0	516	132	76	0.6	989
Greenstone	Yanceyville, N.C.	31	95	40	21[a]		304	76	85	0.3	654
Unconsolidated sand and gravel											
Alluvium	Plymouth, N.H.	23	6.8	1.2	2.6	0.9	17	9.0	5.0	1.2	67
Alluvium	Cave Junction, Oreg.	25	6.4	7.8	5.8	0.2	64	0.8	5.5	0.7	116
Alluvium	Clear Spring, Md.	5.2	36	2.4	1.7	1.1	120	1.0	1.3	0.8	170
Alluvium	Clinton, Iowa	18	44	18	6.0	2.5	144	53	5.0	28.0	318
Glacial outwash	Columbus, Ohio	20	126	43	13	2.1	440	139	8.0	0.2	794
Alluvium	Gila Bend, Ariz.	37	307	82	1,100[a]		327	575	1,820	82	4,340
Alluvium	Pecos, Tex.	43	865	190	738[a]		152	1,910	1,510	468	5,870

SOURCE: White et al., 1963.

[a]Values are sum of sodium and potassium concentrations.
[b]Includes CO_3.

21

weathering of rocks and formation of soil. Another related source involves the concentration of selected elements in vegetation, accompanied by long-term accumulation of the elements in the surface soils and their dissolution by surface runoff. A third source involves wastes from industrial uses.

Minor and trace elements are found in all natural waters, including rain and snow. These elements have not, however, been studied in the careful detail accorded the principal constituents. Although they generally make up less than 1 percent of the TDS, they are of great importance in assessing the suitability of water for livestock and other uses.

Kroner and Kopp (1965), using the National Water Quality Network, summarized available data for trace metals in six major water systems in the United States for the period 1958–1962. Their data are given in Table 6. Barium, copper, and iron appeared most of the time in all systems sampled. Chromium, molybdenum, and nickel also occurred frequently. Other metals generally appeared in less than 20 percent of the samples. Although an element may not have been detected, Kroner and Kopp point out that it possibly was in the original sample. Angino et al. (1969) studied minor and trace element concentrations in the major streams of the lower Kansas River basin (Table 6).

Durum and Haffty (1961) investigated the minor and trace element concentrations in the lower reaches of selected rivers in the United States and Canada. Their data reflected an integrated view of occurrence of the elements in the basins. In all waters examined, they found barium concentrations ranging up to 150 μg/liter and strontium up to about 800 μg/liter. Cesium, rubidium, and lithium were found near the lower levels of detection, suggesting that the alkali metals are strongly absorbed by the soils and rocks upstream. They further reported that

TABLE 6   Concentration Ranges (μg/liter) of Trace Elements in Major Water Supply Systems, 1958–1962

	Ba	Cr	Cu	Fe	Mn	Mo	Na	Pb
Great Lakes	2–50	1–9	1–50	3–90	5–5	1–8	1–10	3–20
Ohio River	20–200	4–10	1–100	5–100	6–30	3–10	3–10	10–30
Mississippi River	20–100	3–20	1–100	6–100	9–50	3–10	3–10	10–50
Missouri River	10–200	8–10	1–50	4–200	–	7–20	6–9	–
Colorado River	20–200	10–30	2–50	10–300	–	5–10	10–30	–
Columbia River	4–60	1–10	1–50	2–90	3–5	1–9	2–10	4–50
Kansas River[a]	–	–	–	–	0–800	–	–	–

SOURCE: Data from Kroner and Kopp, 1965, except where noted.

[a]Angino et al., 1969, in addition to their values for Mn and Na, obtained the following values (μg/liter): Co, 0–300; Zn, 0–600; Li, 0–130; Sr, 10–300; and Ni, 0–120.

zinc and cobalt rarely occur in measured quantities, but that both elements occur in tributaries. Chromium, nickel, copper, lead, titanium, and boron were found in most waters. They concluded that hydrologic and geologic implications are evident in the pattern of occurrence or the absence of a particular minor element.

Soil-forming processes exert dominant controls on the occurrence of minor and trace elements in water. Locally, industrial discharges may contribute a variety of minor and trace elements to livestock water supplies. These principally include cadmium, chromium, copper, lead, manganese, nickel, zinc, mercury, and aluminum. The discharges are also a source of some of the major and secondary constituents as grouped in Table 1.

Sediment

Sediment, defined by the American Society of Civil Engineers as "any material carried in suspension by water which would settle to the bottom if the water lost velocity," is one of the major substances influencing the suitability of water for many uses. Its sources are natural and man-induced erosion of soils and surficial geologic materials and discharge of particulate wastes. Sediment includes both organic and inorganic materials. Because the smaller particles in clay, silt, and sand have chemically active surfaces, many reactions with water occur. Dissolution of cations from soil minerals in the sediment may occur, thus increasing the dissolved solids load of the water. Cation exchange reactions take place and these in turn may cause precipitation of some ions, thereby altering the chemical composition of the water.

Colloidal sediments, because of their much higher exchange capacity, are a major transport medium for the dominant cations calcium, magnesium, and sodium, as well as some of the secondary, minor, and trace constituents. The cation exchange capacity of sediments has been measured by Kennedy (1965). He found the clay fractions in the eastern streams studied to have cation exchange capacities (CEC) of 14–28 meq/100 g; the central and west-central streams, 25–65 meq/100 g; and, in California, 18–65 meq/100 g. He related these figures to the kinds of soils in the various regions. The higher CEC values characteristically occurred in samples containing high proportions of montmorillonite and/or vermiculite. He also found that the CEC of silt ranged from 4 to 30 meq/100 g and that of sand from 0.3 to 13 meq/100 g. From these studies he calculated the ratio of cations absorbed on suspended sediments to those in solution. He found that, when the suspended sediment concentration is high, the ratio approached 1:1 in eastern streams and could be 3:1 in the streams of the West and Southwest.

The mean sediment concentration of rivers has been compiled by Rainwater (1962) and is presented as Figure 5. In areas of good vegetal cover the sediment concentration was found to be as low as 275 mg/liter. In areas of poor vegetal cover and an abundant supply of unconsolidated sediment, the concentration reached 51,000 mg/liter. The larger sediment loads appear to be carried by streams in arid regions. The eastern streams primarily carry sediments in the finer fractions and less sand than the western streams. These sediments can carry important quantities of nutrients, minor elements, and pollutants.

At present, a paucity of meaningful research results exists in regard to the influence of sediment on the chemistry of water. This field needs thorough investigation into both the beneficial and detrimental aspects.

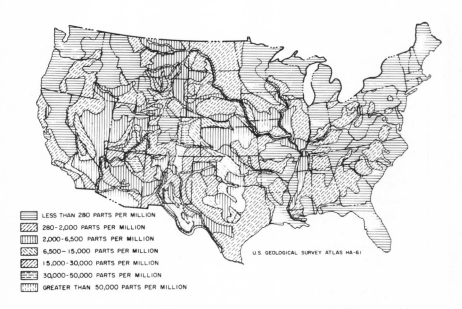

LESS THAN 280 PARTS PER MILLION
280-2,000 PARTS PER MILLION
2,000-6,500 PARTS PER MILLION
6,500-15,000 PARTS PER MILLION
15,000-30,000 PARTS PER MILLION
30,000-50,000 PARTS PER MILLION
GREATER THAN 50,000 PARTS PER MILLION

U.S. GEOLOGICAL SURVEY ATLAS HA-61

FIGURE 5   Sediment concentration of rivers (Rainwater, 1962).

WATER REQUIREMENTS
OF ANIMALS

The water content of animal bodies is relatively constant—68 to 72 percent of the total weight on a fat-free basis. This water level usually cannot change appreciably without severe consequences to the animal. Therefore, the minimal water requirement is influenced by water excreted from the body plus a component for growth in young animals (Robinson and McCance, 1952; Mitchell, 1962).

Water is excreted from the body in urine, feces, sweat, evaporation from the lungs and skin, and productive secretions such as milk and eggs. Anything that influences any of these modes of water loss affects the minimal water requirement of the animal.

The urine contains the soluble products of metabolism that must be eliminated. The amount of urine excreted daily varies according to work performance, external temperature, feed and water consumption, and other factors. The hormone vasopressin (antidiuretic hormone) controls the amount of urine by affecting the reabsorption of water from the kidney tubules and ducts. Under conditions of water scarcity an animal may concentrate its urine to some extent by reabsorbing a greater amount of water than usual, thereby lowering its requirement for water. This capacity for concentration is limited in domestic animals. If an animal consumes excess salt or a high protein diet, the excretion of urine is increased in order to eliminate the salt or the end products of protein metabolism and the water requirement is thereby increased.

The water lost in the feces is variable, depending upon the diet and species. Cattle, for instance, excrete feces high in moisture content as compared with sheep and chickens.

Evaporation from the skin and lungs (insensible water loss) may account for a large part of the water lost from the body, approaching, and in some cases exceeding, that lost in the urine. If the environmental temperature is increased, the water lost by this route is also increased. Water lost through sweating may be considerable particularly among horses, depending on the environmental temperature and activity of the animal.

All of these factors and their interplay make the minimal water requirement difficult to assess from a practical point of view. An additional complication is that the minimal water requirement does not all have to be supplied by drinking water. The animal has available to it the water contained in feeds; the metabolic water formed from the oxidation of nutrients; water liberated by polymerization, dehydration, or synthesis within the body; and preformed water associated with nutrients undergoing oxidation when the energy balance is negative. All of these may vary. The water available from feed will vary with the kind of feed and amount consumed. The metabolic water formed from the oxidation of nutrients may be calculated by the use of factors obtained from equations of oxidation of typical proteins, fats, and carbohydrates. There are 41, 107, and 60 g of water, respectively, formed per 100 g of protein, fat, and carbohydrate oxidized. In fasting animals or in those subsisting on a protein-deficient diet, water may be formed from the destruction of tissue protein. In general, it is assumed that tissue protein is associated with three times its weight in water, so that 3 g of water are released per gram of tissue protein metabolized.

It has been found by careful water balance trials that the requirement of various species for water is a function of their body surface area rather than weight. This implies that the requirements are a function of energy metabolism. Adolph (1933) found that a convenient liberal standard of total water intake is 1 ml/kcal of heat produced. This method of expressing the requirement would automatically include the increased requirement associated with activity. Cattle excrete large amounts of water in the feces; their requirement for water is therefore somewhat higher (1.29–2.05 ml/kcal) than that of other animals.

From a practical point of view, a number of workers have measured water requirement as the amount of water consumed voluntarily under specified conditions. Table 7 presents the water consumption of various classes of livestock under moderate conditions.

Beef cattle, as shown by Winchester and Morris (1956), have daily intakes of water that vary widely depending mostly on ambient temperature and dry matter intake. Thus, heifers and steers of European breeds that weigh approximately 450 kg and eat 10 kg dry feed per day may drink approximately 28, 41, and 66 liters of water daily at 4, 21, and 32 °C, respectively. Hoffman and Self (1972) reported on factors that affected water consumption of feedlot cattle during five summer and four winter trials. They observed that yearling feedlot cattle consumed approximately 50 percent more water in summer than in winter and 8 percent less under shelter in summer. Shelter in winter apparently had no effect. Processing of water with 417 ppm hardness through an ion exchange unit had no significant effect on water consumption or rate of gains.

Dairy cattle, as also illustrated by Winchester and Morris (1956), have water intakes that vary with weight of cow, kg of milk produced daily, fat content of milk, and ambient temperature. They found that a lactating cow at 21 °C weighing approximately 450 kg would consume about 30 liters of water per day plus 2.7 liters of water per kg of milk with 4 percent fat produced. Dairy heifers fed fresh forage and silage may obtain about 20 percent of their water requirement from the feed. Silages vary greatly in their water content. Dairy cattle will suffer more quickly from a lack of water than from a shortage of any other nutrient (NRC, 1971a). Cows producing 40 kg of milk per day may drink up to 110 kg of water when fed dry feeds.

Sheep were studied by Asplund and Pfander (1972) for the effects of water to feed ratios of 1.75:1 and 1:1 at two levels of feed intake during

TABLE 7 Expected Water Consumption of Various Classes of Adult Livestock of Medium Weight in a Temperate Climate

Animal	Liters/Day
Beef cattle	26–66
Dairy cattle	38–110
Horses	30–45
Swine	11–19
Sheep and goats	4–15
Chickens	0.2–0.4
Turkeys	0.4–0.6

a 7-day digestion trial. The rumens of sheep that received the high feed: low water diet rapidly became impacted and only one of four replicates completed the trial. Generally, sheep consume 3 kg of water per kg of dry feed intake (NRC, 1968a). But, many factors may alter this value. They include ambient temperature, activity, age, stage of production, plane of nutrition, composition of feed, and type of pasture. Ewes eating dry feed in winter require 4 liters per head daily before lambing and 6 or more liters per day when nursing (Morrison, 1959).

Swine require 2–2.5 liters of water per kg of dry feed, but voluntary consumption may reach 4–4.5 liters in high ambient temperatures (NRC, 1968b). Mount *et al.* (1971) reported the mean water: feed ratios were between 2.1:1 and 2.7:1 at temperatures between 7 and 22 °C and between 2.8:1 and 5.0:1 at 30 and 33 °C. The range of mean water consumption extended from 0.092 to 0.189 liters per kg body weight per day. Leitch and Thomson (1944) cited studies that demonstrated that a mash prepared with a water-to-meal ratio of 3:1 gave the best results.

Horses need 2–3 liters of water per kg dry ration according to Leitch and Thomson (1944). Morrison (1936) cited data obtained from a horse that gave off 9.4 kg of water vapor when trotting. This amount was nearly twice that given off when walking with the same load and more than three times as much as when resting during the same period. Fonnesbeck (1968) observed *ad libitum* water intake and excretion during three metabolism trials with horses. The horses, at rest at 3–15 °C, were fed diets of hay–grain at approximately maintenance levels. The trials demonstrated that all hay diets resulted in a water-to-feed ratio of 3.6:1, while that of hay–grain diets was 2.9:1; that water intake per kg of dry matter did not correlate to protein content of diets; and that water intake correlated significantly with the ash and cell wall contents of the diet.

Poultry, according to James and Wheeler (1949), consumed more water when protein was increased in the diet. For example, water consumption was higher with meat scrap, fish meal, and dried whey diets than with all-plant diets. Poultry generally consume 2–3 kg of water per kg of dry feed. Sunde (1967) observed that when water was withheld for approximately 36 hours from laying hens at 67 percent production, they dropped to 8 percent production within 5 days and did not return to production of controls until 25–30 days later. Sunde (1971) prepared a table showing that daily water consumption of broilers increased 6.4–211 liters/1,000 birds as they grew from 2 to 35 days of age. Corresponding water intake values for replacement pullets were 5.7–88.5 liters.

NUTRIENTS IN
WATER FOR
LIVESTOCK
AND POULTRY

All the mineral elements essential as dietary nutrients occur to some extent in water (Shirley, 1970). It is generally believed that the availability of elements in water solution is at least equal to that present in solid feeds or dry salt mixes. Shirley and co-workers (1951, 1957) found that radioactive salts of ^{32}P and ^{45}Ca dissolved in water and administered to steers as a drench were absorbed at levels equivalent to those of the isotopes incorporated into forages from fertilizer.

Chapman *et al.* (1962) demonstrated that copper in water solution was much more available to cattle than when incorporated in dry feed. When 12 g of dry copper sulfate were given daily in a gelatin capsule for 12 months, no ill effects were observed; but when this quantity was in aqueous solution as a drench, two treated animals died within 65 days. Numerous isotope studies have demonstrated that minerals in water consumed by animals are readily absorbed, deposited in their tissues, and excreted.

Generally, the elements are in solution, but some may be present in suspended materials. Lawrence (1968) sampled the Chattahoochee River system at six different reservoirs and river and creek inlets. He found about 1, 3, 22, 39, 61, and 68 percent, respectively, of the total calcium, magnesium, zinc, manganese, copper, and iron present in suspended materials. Elements that are a part of suspended materials may not be as available as those in solution to animals drinking the water.

Determinations of the concentration values of most mineral elements in surface waters of the United States during the period 1957–1969 were accumulated in STORET (Systems for Technical Data, 1971). These data include values for the mean, maximum, and minimum concentrations of the nutrient elements (see Table 8). The values obviously include many samples from calcium–magnesium sulfate–chloride and sodium–potassium sulfate–chloride types of water, as well as the more common calcium–magnesium carbonate–bicarbonate types. For this reason, the mean values for sodium, chloride, and sulfate appear somewhat high.

Table 9 gives the estimated average intake of drinking water in liters per day for selected categories of various farm animals. Under the various elements are given three columns of values for illustrative purposes. One column expresses the National Research Council (1966, 1968a,b, 1970a, 1971a,b) daily requirement; the second column gives the approximate mean percentage of that requirement contributed in the water intake each day; and the third column lists the maximum percentage that the daily water intake would supply if the greatest observed concentration of the nutrient were present. No values are presented in Table 9 for percentages of the NRC requirement provided in water when minimum concentrations of nutrients were present, as in nearly all cases they were less than 1 percent.

TABLE 8 Composition of United States Surface Water, 1957-1969 (Collected at 140 Stations)

Substance	Mean	Maximum	Minimum	Number of Determinations
Phosphorus (mg/liter)	0.087	5.0	0.001	1,729
Calcium (mg/liter)	57.1	173.0	11.0	510
Magnesium (mg/liter)	14.3	137.0	8.5	1,143
Sodium (mg/liter)	55.1	7,500.0	0.2	1,801
Potassium (mg/liter)	4.3	370.0	0.06	1,804
Chloride (mg/liter)	478.0	19,000.0	0.0	37,355
Sulfate (mg/liter)	135.9	3,383.0	0.0	30,229
Copper (µg/liter)	13.8	280.0	0.8	1,871
Iron (µg/liter)	43.9	4,600.0	0.10	1,836
Manganese (µg/liter)	29.4	3,230.0	0.20	1,818
Zinc (µg/liter)	51.8	1,183.0	1.0	1,883
Selenium (µg/liter)	0.016	1.0	0.01	234
Iodine[a] (µg/liter)	46.1	336.0	4.0	15
Cobalt[b] (µg/liter)	1.0	5.0	0	720

[a]Dantzman and Breland, 1969.
[b]Durum et al., 1971.

TABLE 9 Mean and Maximum Percentages of Daily Requirements of Nutrient Elements in the Drinking Water of Livestock and Poultry

Animal	Water Intake[a] (liters)	NaCl[c] Req. Daily[b] (g)	Mean (%)	Max. (%)	Calcium Req. Daily (g)	Mean (%)	Max. (%)	Phosphorus Req. Daily (g)	Mean (%)	Max. (%)
Beef cattle (450 kg)										
Nursing cow	60	25	34	4,560	28	12	37	22	<1	1
Finishing steer	60	24	35	4,760	21	16	49	21	<1	1
Dairy cattle (450 kg)										
Lactating cow	90	66	19	2,600	76	7	21	58	<1	1
Growing heifer	60	21	40	5,430	15	23	68	16	<1	2
Maintenance cow	60	21	40	5,430	12	28	86	12	<1	3
Sheep										
Lactating ewe (64 kg)	6	13	7	870	7	5	15	5	<1	1
Fattening lamb (45 kg)	4	10	6	760	3	8	23	3	<1	1
Swine										
Growing (30 kg)	6	4	21	2,800	10	3	10	9	<1	<1
Fattening (60–100 kg)	8	4	28	3,800	17	3	8	14	<1	<1
Lactating sows (200–250 kg)	14	28	7	950	33	2	7	22	<1	<1
Horses (450 kg)										
Medium work	40	90	6	840	14	16	49	14	<1	1
Lactating	50	90	8	1,050	30	10	29	24	<1	1
Poultry										
Chicken (8 wk old)	0.2	0.4	7	1,000	1.0	1	4	0.7	<1	<1
Laying hen (60% production)	0.2	0.5	6	800	3.4	<1	1	0.8	<1	<1
Turkey (8 wk old)	0.2	0.4	7	1,000	1.2	1	1	0.8	<1	<1

31

TABLE 9 (Continued)

Animal	Water Intake[a] (liters)	Magnesium Req. Daily[b] (g)	Mean (%)	Max. (%)	Potassium Req. Daily (g)	Mean (%)	Max. (%)	Sulfur Req. Daily (g)	Mean (%)	Max. (%)
Beef cattle (450 kg)										
Nursing cow	60	14	6	59	90	<1	24	10	27	683
Finishing steer	60	9	10	91	70	<1	31	9	29	719
Dairy cattle (450 kg)										
Lactating cow	90	14	9	88	99	<1	34	20	20	507
Growing heifer	60	9	10	91	70	<1	32	10	27	676
Maintenance cow	60	9	10	91	45	<1	49	6	45	1,127
Sheep										
Lactating ewe (64 kg)	6	1.5	6	68	—	—	—	3	11	270
Fattening lamb (45 kg)	4	1.1	5	49	—	—	—	2	10	250
Swine										
Growing (30 kg)	6	1.0	8	82	8	<1	28	—	—	—
Fattening (60–100 kg)	8	1.0	11	110	9	<1	33	—	—	—
Lactating sows (200–250 kg)	14	2.2	9	90	14	<1	37	—	—	—
Horses (450 kg)										
Medium work	40	13	4	42	30	<1	49	10	18	460
Lactating	50	15	5	46	35	<1	53	10	23	575
Poultry										
Chicken (8 wk old)	0.2	0.05	6	55	—	—	—	—	—	—
Laying hen (60% production)	0.2	—	—	—	—	—	—	—	—	—
Turkey (8 wk old)	0.2	0.05	6	55	—					

32

Beef cattle (450 kg)							
Nursing cow	60	0.99	<1	28	0.15	2	47
Finishing steer	60	0.94	<1	29	0.14	2	51
Dairy cattle (450 kg)							
Lactating cow	90	2.00	<1	21	0.86	1	12
Growing heifer	60	1.00	<1	28	0.43	1	16
Maintenance cow	60	0.64	<1	43	0.27	1	26
Sheep							
Lactating ewe (64 kg)	6	—	—	—	0.25	1	3
Fattening lamb (45 kg)	4	—	—	—	0.18	1	3
Swine							
Growing (30 kg)	6	0.20	<1	14	0.13	<1	5
Fattening (60–100 kg)	8	0.26	<1	14	0.17	<1	5
Lactating sows (200–250 kg)	14	0.44	<1	15	0.28	<1	6
Horses (450 kg)							
Medium work	40	0.39	1	47	—	—	—
Lactating	50	0.39	1	60	—	—	—
Poultry							
Chicken (8 wk old)	0.2	0.008	<1	12	0.005	<1	5
Laying hen (60% production)	0.2	—	—	—	—	—	—
Turkey (8 wk old)	0.2	0.006	<1	15	0.007	<1	3

33

TABLE 9 (Continued)

Animal	Water Intake[a] (liters)	Cobalt			Manganese			Selenium		
		Req. Daily[b] (mg)	Mean (%)	Max. (%)	Req. Daily (mg)	Mean (%)	Max. (%)	Req. Daily (mg)	Mean (%)	Max. (%)
Beef cattle (450 kg)										
Nursing cow	60	0.74	8	40	49	4	395	0.74	1	8
Finishing steer	60	0.71	8	40	47	4	412	0.71	1	8
Dairy cattle (450 kg)										
Lactating cow	90	1.4	6	32	100	3	290	2.0	1	5
Growing heifer	60	0.7	9	43	50	4	388	1.0	1	6
Maintenance cow	60	0.5	12	60	32	6	606	0.6	1	10
Sheep										
Lactating ewe (64 kg)	6	0.18	3	16	—	—	—	0.1	1	6
Fattening lamb (45 kg)	4	0.13	3	15	—	—	—	0.1	1	4
Swine										
Growing (30 kg)	6	—	—	—	50	<1	39	0.25	1	2
Fattening (60–100 kg)	8	—	—	—	66	<1	29	0.33	1	2
Lactating sows (200–250 kg)	14	—	—	—	110	<1	18	0.55	1	3
Horses (450 kg)										
Medium work	40	0.39	10	51	—	—	—	—	—	—
Lactating	50	0.40	12	63	—	—	—	—	—	—
Poultry										
Chicken (8 wk old)	0.2	—	—	—	6	<1	11	0.01	1	2
Laying hen (60% production)	0.2	—	—	—	—	—	—	—	—	—

		Iodine			Copper		
Beef cattle (450 kg)							
Nursing cow	60	1.7	163	1,180	79	1	21
Finishing steer	60	1.6	173	1,260	75	1	22
Dairy cattle (450 kg)							
Lactating cow	90	5.1	81	593	160	1	16
Growing heifer	60	1.6	173	1,260	80	1	21
Maintenance cow	60	1.6	173	1,260	51	2	33
Sheep							
Lactating ewe (64 kg)	6	1.0	28	202	13	1	13
Fattening lamb (45 kg)	4	0.8	23	168	9	1	12
Swine							
Growing (30 kg)	6	—	—	—	15	1	11
Fattening (60–100 kg)	8	—	—	—	20	1	11
Lactating sows (200–250 kg)	14	—	—	—	33	1	12
Horses (450 kg)							
Medium work	40	0.5	369	2,688	78	1	14
Lactating	50	—	—	—	78	1	18
Poultry							
Chicken (8 wk old)	0.2	0.035	26	192	0.4	1	14
Laying hen (60% production)	0.2	0.037	25	182	—	—	—
Turkey (8 wk old)	0.2	—	—	—	0.6	1	9

[a]See discussion on water consumption for sources of these values.
[b]Sources of values generally are the NAS–NRC bulletins on nutrient requirements of various species of livestock and poultry (1966, 1968a,b, 1970a, 1971a,b).
[c]Based on sodium values for water.

Salt as NaCl may be present in significant quantities for some species. Average concentrations of NaCl in surface waters could supply beef cattle with approximately 34 percent of their daily requirement, the lactating dairy cow with 19 percent, growing heifers and cows on maintenance rations with 40 percent, lactating ewes and fattening lambs with 6–7 percent, and growing and fattening swine with 21 and 28 percent, respectively. The lactating sow would receive 7 percent; medium working and lactating horses, 6–8 percent; and broilers, laying hens, and turkeys, 6–7 percent. Average concentration values of NaCl in water are of little practical importance when salt is supplied to livestock on an *ad libitum* basis. If maximum concentrations of NaCl are present, then approximately 8–54 times the daily requirements of the various species would be present. The consequences of such excesses are discussed in the section on salinity (pp. 39–50).

Calcium at average concentrations is present in a sufficient amount to provide approximately 5–8 percent of the requirements for sheep and lactating dairy cows, 10–16 percent for beef cattle and horses, 23–28 percent for growing heifers and maintenance of dairy cows; and 3 percent or less for swine and poultry. Maximum concentrations may supply 15–86 percent of the requirements of beef and dairy cattle, sheep, and horses, but only 1–10 percent for swine and poultry.

Phosphorus at average concentrations in drinking water would provide less than 1 percent of the daily requirements of the six species of farm animals considered, but in a few cases, at maximum concentration, may provide as much as 1–3 percent.

Magnesium in average quantities in drinking water could provide 4–11 percent of the requirements for beef and dairy cattle, sheep, swine, horses, chickens, and turkeys. Maximum concentrations, which seldom occur, would provide 42–110 percent.

Potassium at mean concentrations in drinking water provides less than 1 percent of the requirements of beef and dairy cattle, swine, and horses. At maximum concentrations, 24–53 percent of the requirements are present. Requirements of sheep and poultry for potassium have not been established by the NRC.

Sulfur is provided at approximately 28 percent of beef cattle requirements at average concentrations in drinking water; dairy cattle, 20–45 percent; sheep, 10–11 percent; and horses, 18–23 percent. No daily sulfur requirement values for swine and poultry are available. At maximum concentrations, approximately 2–11 times the daily dietary requirements would be present in normal amounts of water intake; the consequences of these excesses in the form of sulfate are discussed in the section on salinity (pp. 39–50).

Iron at mean concentrations in water provides less than 1 percent of the daily requirements of horses and less for beef and dairy cattle, swine, and poultry. At maximum concentrations, approximately 12–60 percent could be provided. The requirements of sheep for iron have not been established.

Zinc is sufficient at mean concentrations in water to provide 1–2 percent of the requirements of beef and dairy cattle and sheep and less for swine and poultry. At maximum concentrations, 12–51 percent of the requirements of beef and dairy cattle and 3–6 percent of the requirements of sheep, swine, and poultry would be provided. The requirements of horses for zinc have not been established.

Copper at average concentrations may meet 1–2 percent of the daily requirements of six farm species considered. At maximum concentrations in drinking water 9–33 percent of the requirements would be met daily.

Cobalt concentrations reported by Durum *et al.* (1971) were used as typical values for water. The element at the mean level would supply 3–12 percent of the daily requirements of beef and dairy cattle, sheep, and horses. Maximum concentrations of the element in water would provide 15–63 percent of the requirements of these species. Requirements of swine and poultry have not been established.

Manganese at mean concentrations in water could provide 3–6 percent of the dietary requirements of beef and dairy cattle and less than 1 percent of those of swine and poultry. Maximum concentrations would supply approximately 3–6 times the dietary requirements of beef and dairy cattle, 18–39 percent of those of swine, and 11 percent of those of poultry. Requirements of manganese for sheep and horses have not been established.

Selenium at mean concentrations in water provides approximately 1 percent of the dietary requirements of beef and dairy cattle, sheep, swine, and poultry. Maximum levels would supply approximately 1–10 percent of the requirements of these species. Selenium requirements in the diet of horses has not been established. Analyses for selenium concentration in water during the 1930's demonstrated some higher values than those in Table 8. Byers (1935) reported that a drainage water sample in South Dakota contained 1.2 mg/liter of selenium and Byers (1936) found that deposits from drainage ditches in Colorado contained as much as 260 ppm.

Iodine concentrations in 15 rivers and lakes in Florida were analyzed by Dantzman and Breland (1969). Iodine was sufficient at mean concentrations to meet 81–173 percent of the dietary requirements of beef and dairy cattle, 369 percent of those of horses, and 23–38 percent of the requirements of 8-week-old chickens and laying hens and those of sheep. At maximum concentrations, approximately 2 times the requirements of sheep and poultry were available and approximately 6–27 times those of beef and dairy cattle and medium working horses. No requirements have been established for swine, lactating mares, and turkeys. The above Florida data are likely not typical of water in most areas of the United States (Rankama and Sahama, 1950), but little other data are available. Unless water analyses demonstrate that iodine is present, one should probably assume that a given supply of water does not contribute significantly to the needs of farm animals for that element.

SALINITY OF WATER
AS RELATED TO
LIVESTOCK PRODUCTION

Highly saline waters are usually found in the arid or semiarid areas of the world, although they may also occur where seawater contaminates ground sources or in other special circumstances. It has long been known that man or animal restricted to such waters may suffer physiological upset or death. The ions most commonly involved in highly saline waters are calcium, magnesium, sodium, bicarbonate, chloride, and sulfate. Other ions contribute significantly in unusual situations, and some of these may also exert their own specific toxic effects separate from the osmotic effects usually associated with salts. One example is the nitrate ion.

Various methods are used in the literature for expressing the concentration of salts in water. These include parts per million, micrograms per milliliter, and milligrams per liter. For all practical purposes, these are equivalent; all are expressed in this discussion as milligrams per liter (mg/liter).

Hardness is sometimes confused with salinity, but the two are not necessarily correlative. Saline waters containing sodium salts can be very soft if they contain low levels of calcium and magnesium, the divalent cations largely responsible for hardness. Furthermore, because calcium and magnesium are often important components of urinary calculi, hardness has been implicated as a cause of this problem. The literature refuting this is rather extensive.

While research on the effects of saline waters on animals has been somewhat limited, there now has been enough reported to allow for the drafting of preliminary guidelines for their use for livestock. The following discussion reviews this research and sets forth some guidelines based on it.

CATTLE

Early settlers of the West often referred to the salt in waters as "alkali." They attributed a number of animal maladies to it, including the "alkali disease" of livestock now known to be chronic selenium poisoning. Larsen and Bailey (1913), however, found that dairy cows receiving a natural water varying in total mineral content between 4,546 and 7,369 mg/liter for about two years developed neither the "alkali disease" syndrome nor any other abnormalities. When sodium and sulfate ions predominated in the water, the animals suddenly changed from their normal water to the saline solution, refused it for a few days, then drank, and usually had diarrhea. By introducing the cattle to the saline water gradually through a mixing program, even the diarrhea was avoided.

About a decade later, Ramsey (1924) observed that cattle could thrive on water containing 11,400 mg/liter of total salts, and, under some conditions, could live on water with 17,120 mg/liter.

Heller's (1933) work with solutions of various salts and their mixtures led him to conclude that cattle are somewhat less resistant to injury by saline waters than sheep, that 10,000 mg/liter of total salts should be considered the upper limit under which the maintenance of cattle can be expected, and that a lower limit should be used for lactating animals.

Frens (1946) reported studies on the effect on dairy cattle of sodium chloride in their drinking water. A level of 10,000 mg/liter produced no symptoms of toxicity or decrease in milk production during 84 days. However, 15,000 mg/liter caused a loss of appetite, decreased milk production, and, after 12 days, increased water consumption with symptoms of salt poisoning. During the experiment, the chloride content of the blood remained normal.

Embry et al. (1959) reported two experiments with beef cattle. In the first of these, sodium sulfate was added to a control water at levels of 4,000, 7,000, and 10,000 mg/liter. These solutions were offered over a 3-month summer period to heifers (about 300 kg body weight) on a diet of alfalfa hay, shelled corn, and soybean meal full fed with a free-choice mineral mixture. The group fed the highest level of sodium sulfate experienced a severe reduction in water consumption and suffered a loss of weight averaging 0.18 kg/day for 56 days, after which time it was placed on control water without added salt. During the treatment period, scouring was severe. Two of six animals exhibited a period of rapid and difficult respiration and incoordination. Upon removal from the experiment, one revived after being given control water. The other survived without treatment. All animals quickly returned to normal water and feed consumption and weight gain upon being returned to control water. The 4,000 and 7,000 mg/liter of added sodium sulfate caused slightly increased water consumption and reduced free-choice mineral intake, but had no effect on rate of gain or condition of the animals. In the second study, solutions of sodium chloride and a mixture of salts (sodium chloride, magnesium sulfate, and sodium sulfate) of 7,000 and 10,000 mg/liter were offered as the only source of water to steers and heifers for 112 days during the summer. Neither level of the sodium chloride or the mixed salts caused the severely toxic effects found with sodium sulfate, but the higher level of each caused reduced weight gains. Both levels of sodium chloride caused an increase in water intake not found at either level of the mixed salts.

Weeth *et al.* (1960) of the Nevada Agricultural Experiment Station reported that 10,000 mg/liter of sodium chloride in drinking water over a 30-day period caused a 52.8 percent increase in water consumption by heifers and a decrease in blood urea, but no other effects. At 20,000 mg/liter, the salt was toxic, causing severe anorexia, weight loss, lethargy, anhydremia, collapse, and certain other symptoms. Increased urinary urea excretion was found at the 10,000 mg/liter level of salt in the water, and 15,000 mg/liter increased the ratio of urine excretion to water intake from 0.31 during euhydration to 0.82 during salt loading (Weeth and Lesperance, 1965). Later, this group (Weeth *et al.,* 1968) reported that a prompt and distinct diuresis occurred when heifers consumed water containing 5,000 or 6,000 mg/liter of sodium chloride following water deprivation, and that 5,000 mg/liter (Weeth and Hunter, 1971) or even less (Weeth and Caps, 1971) of sodium sulfate added to heifers' drinking water caused reduced water consumption, loss of weight, and increased methemoglobin and sulfhemoglobin levels.

In addition to experimental work, a few observations from the field on the effects of saline waters on cattle have been published. Spafford (1941), in summarizing the results of a survey by questionnaire sent to landowners who had saline waters analyzed by his laboratories, concluded that cattle can thrive on waters containing about 11,400 mg/liter of sodium chloride or 14,250 mg/liter of total salts, but that about 13,800 mg/liter of sodium chloride or 18,500 mg/liter of total salts causes injury. Ballantyne (1957) reported from his observations that cattle offered water containing about 22,000 mg/liter of sodium and magnesium sulfates were reluctant to drink, scoured, staggered, became stiff and blind, and in some cases died. In still another case, he reported that nursing cows in a herd collapsed, were unable to rise, and died when restricted to water containing 19,600 mg/liter of sodium sulfate and 2,000 mg/liter of sodium chloride. He also observed that calves fed water containing 5,400 mg/liter of sodium sulfate showed incoordination and convulsed. In none of these cases did he report the total salts content of the water. Gastler and Olson (1957) also recorded cattle losses where the animals were restricted to excessively saline waters. In three such cases, the total salts contents of the waters involved were 12,000, 20,000, and 32,000 mg/liter.

Based on their observations, officers of the Department of Agriculture and the Governmental Chemical Laboratories (1950) of Australia recommended an upper safe limit for salts in waters of about 7,000 mg/liter for dairy cattle and about 10,000 mg/liter for beef cattle. They suggested that animals can best tolerate highly saline waters when they are on green feeds, if they are offered it for only short periods of time, if they are not lactating, and if they are fully grown.

Sheep

Heller (1933) found that sheep could exist when restricted to waters containing 20,000 mg/liter of magnesium sulfate, or 25,000 mg/liter of sodium chloride or calcium chloride, but not without showing some abnormalities. He found that these animals would not drink highly saline water if allowed a choice between it and water of low salts content.

A series of studies on the effects of saline waters on sheep have been reported by Peirce from Australia. He used merino wethers, a 15-month experimental period, and a diet of wheat and lucerne hays. His control animals were offered rainwater; his experimental animals were offered the same with various amounts of different salts added. Because they usually refused the saline waters when they were first offered, the animals were given a period of a few weeks to become accustomed to them through a program of gradually increasing concentrations. In his first study, Peirce (1957) added only sodium chloride, using levels of 10,000, 15,000, and 20,000 mg/liter. The lowest level produced no effects. The 15,000 mg/liter level reduced feed consumption and the rate of gain in some animals. The highest level reduced feed consumption in all animals and caused a loss of weight, weakness, and listlessness. Also, occasional diarrhea was observed at the two highest levels. While a decrease in wool production occurred for the wethers on the highest level of salt, it was not statistically significant.

Next, Peirce (1959) combined magnesium chloride at various levels up to 5,000 mg/liter with sodium chloride to give about 13,000 mg/liter of total salts. At this level sodium chloride alone or with magnesium chloride up to 1,000 mg/liter had no effect on health, feed intake, or body weight. With magnesium chloride at 2,000 or 5,000 mg/liter, reduced feed consumption adversely affected body weight in some animals and occasional diarrhea was observed. None of the saline waters in this experiment affected wool production.

Again keeping the total salts concentration at about 13,000 mg/liter, Peirce (1960, 1962, 1963) used sodium chloride with various levels of sodium sulfate up to 5,000 mg/liter, calcium chloride up to 3,000 mg/liter, or different mixtures of sodium carbonate and sodium bicarbonate up to 4,100 mg/liter. He found no effects by any of the saline waters on general health, feed consumption, or wool production. Work with "synthetic" waters resembling some underground waters found in Australia with total salts not exceeding 13,000 mg/liter, gave similar results except that a water with a total salts content of 5,000 mg/liter, half of which was sodium bicarbonate, caused a significant decrease in wool production during part of the experimental period (Peirce, 1966).

Using penned (Peirce, 1968a) or grazing (Peirce, 1968b) ewes and their lambs, studies were made of three "synthetic" waters resembling some underground waters of Australia. One had a total salts content of 13,000 mg/liter, mostly sodium chloride; a second had a total salts content of 10,000 mg/liter, again largely sodium chloride; and the third had a total salts content of 5,000 mg/liter, half sodium bicarbonate and almost half sodium chloride. Except for some indication of poorer reproductive performance for the ewes, no adverse effects were observed on health, feed consumption, or wool production of the penned animals. Among the grazing animals, 13,000 mg/liter of total salts decreased body weight gains in lambs and the reproductive rate in the ewes, caused diarrhea, and increased mortality in one of the two experiments. The water with a total salts content of 10,000 mg/liter did not affect the health of lambs, but did decrease their rate of gain and wool production. The alkaline water with 5,000 mg/liter of salts reduced the percentage of ewes that lambed, but had no other effects. Peirce indicated that some of his results suggested that sheep on the more nutritious type of diet would be less susceptible to harm by saline waters than those on nutritionally poor diets.

Officers of the Department of Agriculture and the Governmental Chemical Laboratories (1950) of Australia have recommended about 13,000 mg/liter as the upper safe limit for salts in drinking waters for sheep. As the result of his questionnaire survey, Spafford (1941) stated that sheep apparently will thrive on waters containing about 14,000 mg/liter of total salts but are injured by waters containing 27,000 mg/liter. Reporting on his observations, Ballantyne (1957) stated that lambs were unthrifty and had persistent diarrhea on water containing 3,500 mg/liter of sodium sulfate and 55 mg/liter of nitrate (the concentration of other salts that might have been present was not stated), and that these symptoms disappeared on furnishing the animals better water.

Horses

While no experimental work on the effects of saline waters on horses has been reported in the literature, some individuals have recorded their observations. Ramsey (1924) observed that these animals thrived on waters containing up to about 5,700 mg/liter of soluble salts and could be sustained, if not working too hard, on waters containing up to about 9,100 mg/liter. Spafford (1941) found in his survey that water containing as much as 11,000 mg/liter of sodium chloride or 14,500 mg/liter of total soluble salts had been used for horses for periods of up to 3 months without ill effects and that these animals could be sustained for longer periods on waters with as much as 9,000 mg/liter of sodium chloride or 13,500 mg/liter of total salts. Officers of the Department of Agriculture and Governmental Chemical Laboratories (1950) in Australia have recommended about 6,500 mg/liter as the upper safe limit for salts in waters for horses.

Swine

With swine, Heller (1933) found that 15,000 mg/liter of sodium chloride in the drinking water caused death in small (about 20 kg) animals and some leg stiffness in larger (about 50 kg) ones. Once these animals became accustomed to the taste, 10,000 mg/liter of this salt did not appear particularly injurious.

Work in South Dakota (Embry et al., 1959) included a study with weanling pigs weighing initially about 17 kg. They were fed a growing–finishing diet for 3 months and were watered with a solution containing 2,100, 4,200, or 6,300 mg/liter of a mixture of sodium and magnesium chlorides and sulfates. The added salts caused an increased water intake, but no harmful effects were observed. Berg and Bowland (1960) reported that no harmful effects were observed when sodium chloride, magnesium sulfate, and sodium sulfate were added at levels up to 5,000 mg/liter in the drinking water of growing–finishing pigs.

Ballantyne (1957) reported a field observation in which pigs fed water containing 7,000 mg/liter of sodium chloride scoured and in some cases died. In other field cases, pigs consuming water containing 1,800 mg/liter of sodium chloride and 500 mg/liter of sodium bicarbonate did poorly and those that drank water containing 7,100 mg/liter of sodium sulfate scoured.

The officers of the Department of Agriculture and the Governmental Chemical Laboratories (1950) of Australia recommend an upper safe limit of 4,300 mg/liter of salts in waters for swine.

Poultry

In his work with poultry, Heller (1933) found that waters containing 10,000 mg/liter of sodium chloride did not interfere with normal growth and maintenance of laying hens over a 10-week period, but did delay the onset of egg production. At levels of 15,000 mg/liter of this salt, growth and maintenance were adversely affected, but the same level of magnesium sulfate did no harm. Calcium chloride was even less well-tolerated than was sodium chloride.

Selye (1943) studied the effects on chicks of high levels of sodium chloride in the drinking water. He found a resemblance between avian Bright's disease and experimental intoxication with the salt. Drinking water containing 20,000 mg/liter of sodium chloride was readily consumed by 2-day-old chicks, but within three days all had died, showing signs of diarrhea, generalized tissue edema, and water accumulation in the large serous cavities of the body. Chicks 19 days old that were placed on the experiment and fed water containing 9,000 mg/liter of sodium chloride suffered diarrhea, edema, weakness, gasping respiration, and some deaths during the first 10 days of the trial. Exercise or muscle exertion caused acute cardiac compensation, some dyspnea, and occasional collapse and death. Among those birds that survived for a 20-day period, the edema disappeared, but an increasing development of nephrosclerotic changes appeared to occur. Water containing 3,000 mg/liter of sodium chloride was not toxic to 4-week old chicks.

Later, Kare and Biely (1948) studied the effects of both feed and water with added sodium chloride on chicks. Two-day-old birds on a diet containing 0.182 percent of sodium chloride by analysis and 9,000 mg/liter of added salt in the water experienced some deaths, edema, water in the body cavities, and certain other symptoms. Water with 20,000 mg/liter caused more deaths, fluid in the body cavities, some heart enlargement, lung congestion, and other symptoms. Generalized edema was absent. The chicks exhibited individual differences in susceptibility to the toxicity, and the sodium chloride in the water was about equivalent in toxicity to that in the feed when equal amounts were consumed. A solution containing 18,000 mg/liter of sodium chloride was not toxic when replaced on alternate days with fresh water, but neither was it readily consumed.

Working with 1-day-old turkey poults, Scrivner (1946) offered sodium chloride solutions to the birds while they were on a mash diet containing 0.5 percent of salt. A solution with 20,000 mg/liter caused stupor within 48 hours and the death of all the birds within four days. No edema or ascites were observed. At 5,000 mg/liter the salt caused varying degrees of edema and ascites and death in more than half of the birds in about two weeks. Using sodium bicarbonate solutions, Scrivner found that 1,000 mg/liter was not toxic. Some deaths and edema occurred at 3,000 mg/liter, and these increased as the level of salt was increased to 6,000 mg/liter. A solution containing 1,000 mg/liter of sodium hydroxide caused death and edema in 2 of 31 birds within 13 days, but the remainder survived for 21 days without being obviously affected. Solutions containing 7,500 mg/liter of sodium citrate, iodide, carbonate, or sulfate each caused edema and many deaths. Throughout these studies, birds that survived three or four weeks showed no symptoms of toxicity.

In South Dakota, Krista *et al.* (1961, 1962) also investigated the effects of saline waters on poultry. Using waters with added sodium chloride and diets containing minerals at normal levels, they observed the effects of 4,000, 7,000, and 10,000 mg/liter of the added salt on chicks, laying hens, turkey poults, and ducklings restricted to these waters. Watery droppings were observed in all cases. At the 4,000 mg/liter level, the salt also caused some increased water consumption, decreased feed consumption and growth, and increased mortality. These effects were more pronounced at the higher levels of salt, the 10,000 mg/liter level causing death in all of the turkey poults at 2 weeks, some symptoms of dehydration in the chicks, and decreased egg production in the hens. Experiments with laying hens restricted to water containing 10,000 mg/liter of sodium or magnesium sulfate gave results similar to those for sodium chloride. An interesting fact is that no deaths occurred among laying hens consuming levels as high as 12,000 mg/liter of sodium chloride or sodium sulfate for periods of up to 16 weeks.

The officers of the Department of Agriculture and the Governmental

Chemical Laboratories (1950) of Australia have recommended 2,860 mg/liter as the upper safe limit for salts in drinking waters for poultry.

Rats

While not directly applicable to the problem of excessively saline waters for livestock, the data and comments of Heller and Larwood (1930) and Heller (1932, 1933) on rats should nevertheless be considered here. The more pertinent of these may be summarized as follows:

1. At lower levels of salinity, water consumption increased with increased salt concentration.

2. Salt concentration eventually reached a level at which the rats refused to drink. When thirst finally compelled them to drink, they consumed a large amount at one time and died shortly thereafter.

3. Old animals were more resistant to the effects of the saline waters than were the young.

4. A range of about 15,000–17,000 mg/liter of total salts seemed the maximum that would be tolerated, some adverse effects being observed at even lower levels.

5. The effects produced seemed to be more osmotic in nature than due to any specific ion, except that (a) chloride salts seem less toxic than sulfates; and (b) alkalis were more deleterious than neutral mineral salts, evidently because the osmotic effect was coupled with the harmful effect of high pH.

6. No antagonism between ions was found, and the effects of the various salts seemed additive.

7. Interference with lactation and reproduction was noticeable even before a level of salinity was reached that produced stunted growth and/or death.

8. There appears to be some physiological adjustment in growing rats that in time accommodates their thriving on highly saline waters.

Embry *et al.* (1959) also reported on studies of the effects of saline waters on rats. Sodium chloride (2,925; 5,850; 8,775; and 11,700 mg/liter), sodium sulfate (3,550; 7,100; 10,650; and 14,200 mg/liter), magnesium chloride (2,381; 4,762; 7,143; and 9,524 mg/liter), magnesium sulfate (3,010; 6,019; 9,029; and 12,038 mg/liter), and calcium chloride (2,775; 5,550; 8,325; and 11,100 mg/liter), were each added separately to the drinking water. The sodium chloride increased water intake at the intermediate levels (5,850 and 8,775 mg/liter), as did also the sodium sulfate (7,100 and 10,650 mg/liter). Magnesium salts had little, if any, effect, regardless of level; but calcium chloride decreased water consumption even at the lowest level (2,775 mg/liter). Sodium chloride had little or no effect on growth at concentrations up to 11,700 mg/liter. For the other salts, levels of 10,650 mg/liter of sodium sulfate, 7,100 mg/liter of magnesium chloride, 6,000 mg/liter of magnesium sulfate and 11,000 mg/liter of calcium chloride caused growth depression, but no rats died even at the highest level of any salt during the 50-day experiment. Mild diarrhea was observed among some rats consuming the sulfate salts.

A Guide to the Use of Saline Waters for Livestock

The literature reveals a rather wide variation in experimental results concerning the effects of saline waters on livestock. This variation indicates the need to take into account a number of factors in evaluating these waters for livestock use. They include the kind, age, and sex of the animals; whether they are pregnant or lactating; the intensity of the work or exercise they perform; climatic conditions; type of diet and moisture content; level of production; amount of minerals in the diet; type and level of salts in the water; access to other sources of water; and whether or not the animals have been adapted to the water. The weight that each of these factors must be given in any particular situation is still largely a matter of judgment, but there seems little doubt that the total soluble salts content is the single most reliable parameter by which saline waters can be evaluated for livestock use. Table 10 is based upon this parameter.

It is not the purpose of this guide to recommend the use of highly saline waters for livestock. Whenever possible, drinking water should have a low mineral content. However, in the many cases where circumstances are such that highly saline waters are all that is readily available, a guide to their use should be available for livestock producers.

When using Table 10, the various factors listed above should be given some consideration. In addition, the following points should be taken into account:

1. Whenever water has a total salts content of more than 3,000 mg/liter, alkalinities and nitrates should also be considered. Alkalinities of 2,000 mg/liter expressed as $CaCO_3$ detract from the suitability of the waters. Hydroxides are more harmful than carbonates, which in turn are more harmful than bicarbonates.

2. If animals are offered two sources of water, one highly saline and the other not, they will not drink the highly saline water.

3. Animals can consume water of very high salinity for a few days without being harmed if they are then given water of low soluble salt content.

4. As the soluble salts content of water increases, intake usually increases, except for water of extremely high saline content that the animals refuse to drink.

5. Abrupt change from water of low salinity to that of high salinity will probably cause more problems than gradual change.

6. Depressed water intake is very likely to be accompanied by depressed feed intake. Thus, animals being fed for a high rate of gain or

TABLE 10 A Guide to the Use of Saline Waters for Livestock and Poultry

Total Soluble Salts Content of Waters (mg/liter)	Comment
Less than 1,000	These waters have a relatively low level of salinity and should present no serious burden to any class of livestock or poultry.
1,000–2,999	These waters should be satisfactory for all classes of livestock and poultry. They may cause temporary and mild diarrhea in livestock not accustomed to them or watery droppings in poultry (especially at the higher levels), but should not affect their health or performance.
3,000–4,999	These waters should be satisfactory for livestock, although they might very possibly cause temporary diarrhea or be refused at first by animals not accustomed to them. They are poor waters for poultry, often causing watery feces and (at the higher levels of salinity) increased mortality and decreased growth, especially in turkeys.
5,000–6,999	These waters can be used with reasonable safety for dairy and beef cattle, sheep, swine, and horses. It may be well to avoid the use of those approaching the higher levels for pregnant or lactating animals. They are not acceptable waters for poultry, almost always causing some type of problem, especially near the upper limit, where reduced growth and production or increased mortality will probably occur.
7,000–10,000	These waters are unfit for poultry and probably for swine. Considerable risk may exist in using them for pregnant or lactating cows, horses, sheep, the young of these species, or for any animals subjected to heavy heat stress or water loss. In general, their use should be avoided, although older ruminants, horses, and even poultry and swine may subsist on them for long periods of time under conditions of low stress.
More than 10,000	The risks with these highly saline waters are so great that they cannot be recommended for use under any conditions.

production may be expected to show deleterious effects from waters of lower salts content than animals on a maintenance regimen.

7. While water should normally not be relied upon as a source of essential inorganic elements (as discussed elsewhere in this report), highly saline waters may in some situations furnish enough of these to be considered in calculating mineral additions to the diet. Furthermore, the salt content of the diet may contribute to the toxicity of saline waters. This is of particular concern when salt additions to diets are used to control feed intake.

TOXIC ELEMENTS AND
SUBSTANCES IN WATER

Whether from natural or human sources, water occasionally contains elements or substances in toxic amounts. Unfortunately, only limited information is available on experimentally determined toxic levels of various substances in water for livestock and poultry. Good judgment is necessary when applying information reported in the literature to practical situations. So many conditions are involved in determining whether or not certain levels of a toxicant will cause harm that no single concentration can be accepted as dangerous in all situations.

Following are some examples of the problems involved in assigning toxic levels to waterborne substances. Toxic substances in water may be either a part of the suspended solids, in solution, or distributed between the two; their availability in these phases may differ considerably during the digestion process of animals. Different valences and chemical forms of the elements frequently result in different toxicities to the animals consuming them. Short-term intake of a toxic substance may have no observable effects, while long-term consumption may result in serious harm. Different species of animals may react differently to a toxic substance, and the young and healthy may not respond in the same way as mature or unthrifty animals. The rate of consumption may be involved. Antagonistic or synergistic substances present in either water or feed, or additional toxic material in the feed, may at times be significant factors. Finally, intake of toxic substances that cause no measurable effect on growth, production, or reproduction may cause subcellular damage that expresses itself as increased susceptibility to disease or to parasitic invasion. Some elements may not be toxic to livestock that consume them, but may accumulate in meat, milk, or eggs at concentrations harmful or objectionable to persons who eat them.

It should be pointed out that water sources, especially those from the surface or shallow wells, are subject to sudden changes in composition from natural or human causes.

A number of elements found in water seldom offer any problem to livestock because they do not occur at high levels in soluble form or because they are toxic only in excessive concentrations. Examples of these are iron, aluminum, beryllium, boron, chromium, cobalt, copper, iodide, manganese, molybdenum, and zinc. Also, these elements do not seem to accumulate in meat, milk, or eggs to the extent that they would constitute a problem in livestock drinking waters under any but the most unusual conditions.

On the other hand, elements such as lead, mercury, and cadmium must be considered actual or potential problems, because they occasionally are found in waters at toxic concentrations or may accumulate in meat, milk, or eggs at levels unfit for human consumption.

The toxic levels of inorganic elements in feeds for poultry were included in the NRC report on the *Nutrient Requirements of Poultry* (1971b). This report also listed toxic concentrations of sodium sulfate and sodium chloride in water for laying hens and sodium nitrate in water for turkeys. The maximum levels of cobalt, copper, zinc, molybdenum, fluoride, and selenium recommended for dairy rations were included in Table 3 of the NRC report on *Nutrient Requirements of Dairy Cattle* (1971a). Tables 11 and 12 briefly summarize data for livestock and poultry from the literature on the toxic effects, if any, of various elements and substances in water and feeds. In Table 13 are presented concentrations of potentially toxic substances that probably should not be exceeded in the drinking water of livestock and poultry. Many dietary, physiological, and environmental factors make it difficult to determine precise concentrations that will not be harmful in drinking water. Nevertheless, evaluations of the literature were made in regard to what concentrations should be safe in maintaining the health of the animals and the quality of their products for human consumption. Many of these limits (Table 13) are the same as those recommended by the panel on water for uses in agriculture (NAS–NAE, 1972).

In addition to the summaries on toxic substances compiled in Tables 11 and 12, the following discussion contains certain other information that may be useful in arriving at decisions on the use of waters containing potential toxicants.

Arsenic, well-known for its toxicity and tendency to accumulate under some conditions in certain tissues or fluids, has also been accused of being a carcinogen. Frost (1967), however, has refuted this charge and has also stated that no evidence exists that 10 ppm of arsenic in the diet is toxic to any animal, suggesting that it may be less toxic than we sometimes give it credit for being. Arsenic occurs in some detergents (Pattison, 1970), which may be one source, although probably of little significance, for the element in waters. Arsenic in soils is unavailable to plants (Olson *et al.*, 1940), possibly because of its long-known fixation by iron hydroxide (Boswell and Dickson, 1918). Absorption of this type may account for its unusually low concentration in natural waters. The chemical form of this element will, of course, influence its toxicity and may also affect the manner in which it accumulates in body tissues. Pentavalent arsenic is regarded as considerably less toxic than the trivalent form (Byron *et al.*, 1965). The biological methylation of arsenic needs further study to evaluate its significance in regards to toxic effects on tissue accumulation.

Peoples (1964) fed arsenic acid at levels up to 1.25 mg/kg body wt/ day for 8 weeks to lactating cows. This is equivalent to an intake of 60 liters of water containing 5.5 mg/liter of arsenic daily by a 500 kg animal. His results indicated that this form of arsenic is rapidly absorbed and excreted in the urine. Thus, tissue storage was minimal. At no level of the added arsenic did an increased arsenic content of the milk occur, and no toxicity was observed. However, when Bucy *et al.* (1955) fed 0.05, 0.1, 0.2, and 0.4 percent arsenic as potassium arsenite, arsanilic acid, or 3-nitro-4-hydroxyphenylarsonic acid in practical rations to sheep, they observed significant increases of arsenic in the liver, kidney, and muscle and pathological changes in the liver and kidney.

United States Public Health Service standards for drinking water (United States Department of Health, Education, and Welfare, 1962b) list 0.05 mg/liter of arsenic as the upper allowable limit for man, but McKee and Wolf (1963) suggest 1.0 mg/liter as the upper limit for livestock.

Cadmium contamination of groundwaters has been reported by Lieber and Welsch (1954). The element has been found in groundwater contaminated by electroplating wastes at 3.2 mg/liter and in mine waters at more than 1,000 mg/liter (McKee and Wolf, 1963). Decker *et al.* (1958) demonstrated that 50 mg/liter of cadmium as $CdCl_2$ in drinking

TABLE 11 Summary of Effects of Toxic Levels of Elements and Substances in Water for Livestock, Poultry, and Experimental Animals

Element	Concentration	Species	Effects	Reference
Antimony (Sb)	5 mg/liter	mice	decreased growth and longevity of females	Schroeder et al., 1968a
Arsenic (As) (see discussion in text)	5 mg/liter	mice	nontoxic, but accumulated in some tissues	Schroeder and Balassa, 1967
	5 mg/liter	mice	survived through three generations	Schroeder and Mitchener, 1971b
	1 mg/liter	guinea pigs	increased thyroidal colloid	Bocconi and Bonessa, 1953
Barium (Ba)	2 mg/liter	man	probably not toxic	Stokinger and Woodward, 1958
Beryllium (Be)	28.5 mg/liter	goldfish, minnows, and snails	not very toxic	Pomelee, 1953
Boron (B)	2 mg/liter	lambs	stated to be excessive	Plotnikov, 1960
Cadmium (Cd) (see discussion in text)	10 mg/liter	mice	second generation didn't survive	Schroeder and Mitchener, 1971b
	5 mg/liter	rats	reduced longevity	Schroeder et al., 1963b
	5 mg/liter	mice	reduced longevity	Schroeder et al., 1963a
	0.1 mg/liter	rabbits	in 6 mo kidney epithelium swollen	Grushko et al., 1951
	0–5 mg/liter	dogs	tissue concentration in proportion to intake	Byerrum et al., 1960
Calcium (Ca) (see section on nutrients)				
Chloride (Cl) (see section on salinity)				
Chromium (Cr)	500 mg/liter	rats and rabbits	maximum nontoxic level based on growth	Gross and Heller, 1946
	25 mg Cr^{6+}/liter	rats	decreased water intake; 9 times more present in tissue than when Cr^{3+} substituted	MacKenzie et al., 1958
	5 mg/liter	livestock	not affected adversely	McKee and Wolf, 1963
	10 mg/liter	man	some nausea over 15-day period	McKee and Wolf, 1963
Cobalt (Co)	1.1 mg/day/kg body wt in drench	calves	symptoms similar to Co deficiency	Keener et al., 1949
	160 mg/day in drench	sheep	depressed appetite; weight losses	Becker and Smith, 1951
	5 mg/liter	rats	increased mortality with 11 ppm Se in diet	Moxon and Rhian, 1943

Substance	Concentration	Animal	Effect	Reference
Copper (Cu) (see discussion in text)	12 g CuSO$_4$ · 5H$_2$O/day in drench	steers	body weight decrease; fatal in 65 days	Chapman et al., 1962
	625 mg/liter	turkeys	decreased feed and water intake; fatal	Hinshaw and Lloyd, 1931
Cyanide (CN)	103 mg HCN/liter	cows and ducks	fatal	Clough, 1934
Dissolved solids	5,140 mg/liter	pigs	safe upper limit	Anon., 1950
	12,000 mg/liter	dairy cows	reduced milk yields	Anon., 1950
	8,600 mg/liter	beef cows	safe upper limit	Anon., 1950
	7,000 mg/liter	cattle	loss of body weight	Anon., 1959
	15,600 mg/liter	dry sheep	safe upper limit	Anon., 1950
	3,430 mg/liter	poultry	safe upper limit	Anon., 1950
Fluoride (F⁻) (see discussion in text)	100 mg/liter	calves	decreased feed intake, growth and Ca absorption; bones decalcified	Ramberg et al., 1970
	11.8 mg/liter	cattle	mottled teeth	Ockerse, 1943
	5–10 mg/liter	sheep	mottled teeth	Peirce, 1952
	10 mg/liter	sheep	decreased wool production	Peirce, 1959
	20 mg/liter	sheep	decreased health; severe teeth mottling	Peirce, 1959
	6–10 mg/liter	hogs	severe mottling of teeth	McKee and Wolf, 1963
	10 mg/liter	mice	no innate toxicity	Schroeder et al., 1968a
Germanium (Ge)	5 mg/liter	mice	reduced life span of males; accumulation in tissues	Schroeder and Balassa, 1967
	5 mg/liter	rats	no effect on longevity	Schroeder et al., 1968b; Arrington, 1973
Iodide (I)	(no published data for toxicity in water; approximately same as in feed)			
Iron (Fe)	17 mg/liter	cattle	in pasture irrigation water; scouring; decreased milk production and body weight	Coup and Campbell, 1964
Lead (Pb) (see discussion in text)	100 mg/liter	calf	died after 4 mo of drinking Pb(NO$_3$)$_2$	Allcroft, 1951
	25 mg/liter	rats and mice	reproduction ceases in second generation	Schroeder and Mitchener, 1971b
Lithium (Li)	5 mg/liter	rats	hypertension and higher death rates due to infections	Schroeder et al., 1965
Magnesium (Mg) (see section on salinity)	5 mg/liter	man	limit for drinking and cooking	Hibbard, 1934

55

TABLE 11 (Continued)

Element	Concentration	Species	Effects	Reference
Manganese (Mn)	500 mg/liter	cattle	reduced liver Fe but no toxicity	Weeth, 1962
Mercury (Hg)	75–300 mg/liter	man	fatal	Smith, 1944
(see discussion in text)	30 µg/liter	man	readily absorbed into tissues	Aberg et al., 1969
Molybdenum (Mo)	5 mg/liter	rats	increased mortality with 11 ppm Se in diet	Moxon and Rhian, 1943
	10 mg/liter	mice	third generation failed to survive	Schroeder and Mitchener, 1971b
Nickel (Ni)	5 mg/liter	rats	reduced litter size through three generations	Schroeder and Mitchener, 1971b
Niobium (Nb)	5 mg/liter	mice	decreased growth and longevity; increased fatty degeneration	Schroeder et al., 1968a
Nitrate (NO₃⁻)	667 mg N/liter	lambs	methemoglobin 5% of hemoglobin in 84-day trial	Seerley et al., 1965
(see discussion in text)	1,000 mg N/liter	sheep	16% of hemoglobin converted to methemoglobin	Seerley et al., 1965
	300 mg N/liter	sheep	no detrimental effect	Seerley et al., 1965
	300 mg N/liter	pigs	no effect on weight gain or reproduction during two gestation periods	Seerley et al., 1965
	90 mg N/liter	cattle	no observed harm with range cattle	Hale, 1973
	300 mg N/liter	chicks	no effect on feed and water intake and growth	Adams et al., 1966
	300 mg N/liter	laying hens	no effect on egg production or quality	Adams et al., 1966
450 mg N/liter	450 mg N/liter	turkeys	no effect on meat color over 24 weeks	Mugler, 1970
	4,158 mg N/liter	guinea pigs	67% fetal loss	Sleight and Atallah, 1968
Nitrite (NO₂⁻)	100 mg N/liter	pigs	small increase in methemoglobin	Seerley et al., 1965
(see discussion in text)	150 mg N/liter	cockerels and poults	no effect on growth, feed efficiency, methemoglobin, or thyroid weight to 4 weeks of age	Kienholz et al., 1966
	200 mg N/liter	chicks	decreased growth and liver vitamin A	Adams et al., 1966
	200 mg N/liter	laying hens	no effect on egg production and quality	Adams et al., 1966
	200 mg N/liter	turkeys	decreased feed intake and liver vitamin A	Adams et al., 1966

56

Substance	Concentration	Organism	Effect	Reference
	20 mg N/liter	rats	decreases life span slightly over 3 generations	Druckrey et al., 1963
Potassium (K) (excesses not likely to occur in water; see section on nutrients)				
Selenium (Se) (see discussion in text)				
Se^{4+}	2–3 mg/liter	rats	decreased growth and early death in males	Schroeder and Mitchener, 1971a
Se^{6+}	2–3 mg/liter	rats	no effect on growth and longevity, but tumorigenic	Schroeder and Mitchener, 1971a
Se^{6+}	3 mg/liter	mice	third generation failed to survive	Schroeder and Mitchener, 1971b
Sodium Chloride (NaCl) (see section on salinity)				
Sulfate (see section on salinity)				
(SO_4^-) as S	100 mg/liter	cattle	lost weight; decreased water intake 35%, feed intake by 30%, and creatinine excretion by 12%	Weeth and Hunter, 1971
	3,590 mg/liter	cattle	weakened and died; bones decalcified	Allison, 1930
	2,700 mg/liter as Na_2SO_4	laying hen	reduced egg production	Krista et al., 1961
Tellurium (Te)	2 mg/liter	rats	nontoxic	Schroeder and Mitchener, 1971b
Tin (Sn)	5 mg/liter	mice and rats	decreased longevity in females; fatty degeneration of liver; vascular changes in kidney	Schroeder et al., 1968b
Titanium (Ti)	5 mg/liter	mice	no toxic effect	Schroeder and Balassa, 1967
Vanadium (V)	5 mg/liter	rats	few survived to third generation	Schroeder and Mitchener, 1971b
	5 mg/liter	rats	increased mortality when 11 ppm Se in diet	Moxon and Rhian, 1943
Zirconium (Zr)	5 mg/liter	mice and rats	no effect on longevity	Schroeder and Balassa, 1967
Zinc (Zn)	5 mg/liter	mice	slightly toxic; tissue accumulation	Schroeder et al., 1968a
	675 mg/liter	human	emetic	Sandstead et al., 1970
	5 mg/liter	rats	increased mortality with 11 ppm Se in diet	Moxon and Rhian, 1943
	2,320 mg/liter	hens	decreased water consumption; egg production stopped after 3 days; body weight decreased	Sturkie, 1956

TABLE 12 Summary of Toxic Effects of Minerals, Nitrates, and Nitrites in Livestock and Poultry Feed

Element	Quantity in Diet	Species	Effects	Reference
Aluminum (Al)	2,200 ppm	chick	rickets	Deobold and Elvehjem, 1935
	500 ppm	immature chicken	reduced growth	Storer and Nelson, 1968
	1,000 ppm	immature chicken	reduced growth (1.6% Al_2O_3 no effect)	Storer and Nelson, 1968
Arsenic (As)	1.25 mg/kg body wt	cows	no effect over 8-week period	Peoples, 1964
	2,000 ppm	sheep	convulsions; went off feed; death	Bucy et al., 1955
Barium (Ba)	300 ppm	chick	depressed weight gain	Taucins et al., 1969
	6,000 ppm	chick	death	Taucins et al., 1969
Bromide (Br⁻)	0.5 ppm	rat	growth normal	Winnek and Smith, 1937
	5,000 ppm	immature chicken	reduced growth	Doberenz et al., 1965
Cadmium (Cd)	3 g/day	cows	reduced milk production	Miller et al., 1967
	160 ppm	calves	reduced feed intake	Powell et al., 1964
	60 ppm	lambs	reduced feed intake and growth	Doyle et al., 1974
	100 ppm	immature chicken	reduced growth	Hill et al., 1964
	25 ppm	immature chicken	reduced growth when deficient in Cu and Fe	Hill et al., 1963
Chloride (Cl⁻)	20 ppm	immature turkey	reduced growth	Supplee, 1961
	15,000 ppm	immature chicken	reduced growth but effect largely overcome by adding Na and K	Nesheim et al., 1964
Chromium (Cr)	300 ppm (K_2CrO_4)	immature chicken	reduced growth	Kunishisa et al., 1966
	100 ppm	chick	no effect over 21-day period	Romoser et al., 1961
Cobalt (Co)	4–11 mg/kg body wt	cattle	depressed appetite and weight loss; anemia; death	Becker and Smith, 1951
	40–60 mg/kg body wt in drench	sheep	fatal at high levels within 16 h	Andrews, 1965
	5 ppm	immature chicken	reduced growth	Turk and Kratzer, 1960
	50 ppm	immature chicken	mortality	Turk and Kratzer, 1960
Copper (Cu)	40 ppm	sheep	toxic	Ross, 1964
	80 ppm	sheep	toxic; brain lesions; death	Doherty et al., 1969
	38 mg/day	sheep	toxic	Hemingway and MacPherson, 1967
	115 ppm	calves	toxic	Shand and Lewis, 1957

Element	Level	Animal	Effect	Reference
Fluoride (F⁻)	324 ppm	immature chicken	reduced growth	Mayo et al., 1956
	500 ppm	chick	depressed weight gain	Taucins et al., 1969
	1,000 ppm	chick	death	Taucins et al., 1969
	676 ppm	immature turkey	reduced growth	Vohra and Kratzer, 1968
	50 ppm	immature turkey	(purified diet) reduced growth	Waibel et al., 1964
	800 ppm	immature turkey	(practical diet) reduced growth	Waibel et al., 1964
	70–100 ppm	cows	decreased reproduction	Hobbs and Merriman, 1962
	1,000 ppm	immature chicken	reduced growth	Doberenz et al., 1965
Germanium (Ge)	(injections)	rats	low order of toxicity	Rosenfeld and Wallace, 1953
Iodide (I)	50 mg KI/day	sheep	reduced reproduction	Malan et al., 1935
	2,500 ppm	swine	no effect on dam or offspring	Arrington et al., 1965
	2,500 ppm	hamsters	reduced weaning weight	Arrington et al., 1965
	250–1,000 ppm	rabbits	increased mortality of newborn when fed diets for 2–5 days	Arrington et al., 1965
	625–5,000 ppm	laying hen	reduced egg production, egg size, and hatchability; if deleted I production becomes normal	Arrington et al., 1967
Iron (Fe)	210 and 280 ppm	sheep	diarrhea and death	Lawlor et al., 1965
	400 ppm	feedlot calves	depressed gain over 84-day period	Standish et al., 1969
	4,000 ppm	pig	reduced gain; decrease in serum P	O'Donovan et al., 1963
	4,500 ppm	chick	rickets	Deobold and Elvehjem, 1935
Lead (Pb)	1 mg/kg body wt	ewe	abortion and death	Allcroft and Blaxter, 1950
	200–400 mg/kg body wt	calves	death in a few cases	Allcroft and Blaxter, 1950
	3 ppm	cattle	no Pb retained in tissues	Blaxter, 1950
	1,000 ppm	chick	growth depression	Damron et al., 1969
	100 µg as PbNO₃	mice	injected daily for 30 days, less resistance to salmonella typhimurium	Hemphill et al., 1971
Magnesium (Mg)	6,000 ppm	immature chicken	reduced both growth and bone mineralization	Chicco et al., 1967
Manganese (Mn)	6,400 ppm	immature chicken	reduced growth; increased mortality	Nugara and Edwards, 1963
	45 ppm	lambs	decreased Fe absorption; anemia	Hartman et al., 1955
	100 ppm	beef cattle	depressed serum Mg; no effect on Ca, Fe, K, and glucose	Fain et al., 1952
	50–125 ppm	baby pigs	hemoglobin reduced	Matrone et al., 1959
	500 ppm	pigs	depressed appetite and growth	Grummer et al., 1950
	4,800 ppm	immature turkey	reduced growth	Vohra and Kratzer, 1968

TABLE 12 (Continued)

Element	Quantity in Diet	Species	Effects	Reference
Mercury (Hg) (see discussion in text)				
Molybdenum (Mo)	1,000 ppm	pigs	rapid absorption and excretion; nontoxic	Davis, 1950
	300–350 ppm	young bull calves	stiff joints; diarrhea	Thomas and Moss, 1951
	200 ppm	immature chicken	reduced growth	Arthur et al., 1958
	500 ppm	immature chicken	reduced growth; mortality	Davies et al., 1960
	500 ppm	laying hen	reduced egg production and hatchability	LePore and Miller, 1965
Nickel (Ni)	250–1,000 ppm	calves	poor gains or loss in weight	O'Dell et al., 1970
	1,100–1,600 ppm	mice	reduced growth; no effect on body weight of adults; 1,600 ppm decreased number weaned	Weber and Reid, 1969
Nitrate (NO$_3^-$)	700 ppm	chicks	reduced growth and N retention	Weber and Reid, 1968
	83 mg N/kg body wt	sheep	splenomegaly	Miyazaki, 1967
	99 mg N/kg body wt + hay	sheep	1 death	Buchman et al., 1968
	112 mg N/kg body wt in drench	sheep	>5 g methemoglobin/100 ml blood in about 8 h	Emerick et al., 1965
	112 mg N/kg body wt	sheep	30% weight reduction	Miyazaki, 1967
	0.5% N	sheep	46% decrease in liver vitamin A after 56 days	Goodrich et al., 1964
	1.0% N	sheep	abortion; death	Davison et al., 1965
	6,700 ppm	sheep	depressed gain	Sokolowski, 1966
	350 mg N/kg body wt by injection	sheep	diuresis	Pfander et al., 1957
	45 mg N/kg body wt	dairy cows	death in 3–13 days	Simon et al., 1959
	0.23% N	dairy cows	no effect	Jones et al., 1966
	63 mg N/kg body wt	cattle	death on second day	Simon et al., 1959
	100 mg N/kg body wt	heifers	1 abortion	Davison et al., 1964
	150 mg N/kg body wt	heifers	low conception; 2 abortions; 2 deaths	Davison et al., 1964
	150 mg N/kg body wt	heifers	no effect on reproduction; cyanosis	Winter and Hokanson, 1964
	74 mg N/kg body wt	cattle	MLD$_{50}$	Bradley et al., 1940

60

Amount	Animal	Effect	Reference
450 mg N/kg body wt	cattle	LD$_{50}$[a]; no abortion; no decreased weight or milk production	Crawford et al., 1966
0.16% N	cattle	decreased feed intake and weight gain	Weichenthal et al., 1963
0.54% N	swine	growth depression; no effect on reproduction	Tollett et al., 1960
Nitrite (NO$_2^-$) (see discussion in text)			
0.9% N	swine	growth decrease	Hutagalung et al., 1968
15 mg N/kg body wt	sheep	death in 3 h	Pfander et al., 1957
34 mg N/kg body wt	sheep	calculated to be lethal	Burden, 1968
66 mg N/kg body wt	sheep	death in 2–3 h	Sinclair and Jones, 1967
13 mg N/kg body wt	sheep	death in 2 h	Diven et al., 1964
80 mg N/kg body wt	sheep	death	Lewis, 1951a,b
80 mg N/kg body wt	sheep	reduced liver Vitamin A	Holst et al., 1961
20 mg N/kg body wt	cattle	LD$_{50}$[a]	Stormoken, 1953
18 mg N/kg body wt	pigs	calculated lethal	Burden, 1968
21 mg N/kg body wt	pigs	LD$_{50}$[a]	Nelson, 1966
23 mg N/kg body wt	pigs	death	Winks et al., 1950
0.09% N	swine	decreased weight gains; increase in methemoglobin	Hutagalung et al., 1968
0.6 mg N/kg body wt	rabbit	hypotension; tachycardia	Iacovoni et al., 1968
18 mg N/kg body wt	rabbit	calculated lethal dose	Burden, 1968
121 ppm N	immature turkey	reduced growth; mortality	Sunde, 1964
365 ppm N	chicks	decreased vitamin A in liver and thyroid hypertrophy	Sell and Roberts, 1963
Rubidium (Rb)			
200–1,000 ppm	rats	poor survival	Glendening et al., 1956
Selenium (Se)			
0.25–0.40 mg/kg body wt as selenite	cattle	death of some animals in 8 weeks	Maag and Glenn, 1967
37 mg Na$_2$SeO$_4$/day	ewes	deaths after 80 days	Maag and Glenn, 1967
10 ppm	sows	unsatisfactory reproduction; weanling pigs had decreased gains, hoof lesions, loss of hair	Wahlstrom and Olson, 1959a,b
10 ppm	laying hen	reduced hatchability of eggs	Moxon and Wilson, 1944
Na$_2$SeO$_3$/kg body wt	male rats	7.5 mg is LD$_{50}$[a]	Pletnikova, 1970
	female rats	11.9 mg is LD$_{50}$[a]	Pletnikova, 1970
	guinea pigs	5.1 mg is LD$_{50}$[a]	Pletnikova, 1970
Silver (Ag)			
200 ppm	immature chicken	reduced growth	Hill et al., 1964
Strontium (Sr)			
5,000 ppm	chick	depressed weight gain	Taucins et al., 1969

61

TABLE 12 (Continued)

Element	Quantity in Diet	Species	Effects	Reference
Vanadium (V)	20 mg/kg body wt	calves	diarrhea and emaciation	Platnow and Abbey, 1968
	30 ppm	chick	depressed weight gain	Romoser et al., 1961
	200 ppm	chick	caused death	Romoser et al., 1961
	10–100 ppm	laying hens	reduced growth with increasing intakes	Berg et al., 1963
	30–40 ppm	laying hens	depressed albumen quality	Berg et al., 1963
	47 ppm	chick	depressed growth	Nelson et al., 1962
	30–40 ppm	laying hens	decreased egg production	Berg et al., 1963
	50 ppm	laying hens	decreased hatchability of eggs	Berg et al., 1963
	25 ppm	chicks	decreased growth with deaths; storage in bone and kidney	Hathcock et al., 1964
Zinc (Zn)	900 ppm	cattle	reduced gains	Ott et al., 1966b,d
	1,000 ppm	lambs	reduced gains	Ott et al., 1966a,c
	20 ppm	cattle	decreased in vitro digestion of cellulose by rumen microbes	Martinez and Church, 1970
	1,000 ppm	weanling pigs	no effect	Brink et al., 1959
	2,000–8,000 ppm	weanling pigs	decreased wt gains; gastritis; hemorrhage; deaths	Brink et al., 1959
	3,000 ppm	broilers	toxic	Pensack and Klussendorf, 1956
	3,000 ppm	laying hens	toxic	Pensack and Klussendorf, 1956
	15,000 ppm	immature chicken	reduced growth (with both $ZnCO_3$ and $ZnSO_4$)	Roberson and Schaible, 1960
	3,000 ppm	immature chicken	reduced growth	Johnson et al., 1962
	4,000 ppm	immature turkey	reduced growth	Vohra and Kratzer, 1968

[a]LD_{50} is a dose that causes death in 50% of the animals involved.

62

water of rats during a period of 320 days caused stunted growth and nearly a 50 percent decrease in hemoglobin. Doyle and Pfander (1974) gave lambs a choice of water containing 0, 3, 5, 8, 12, and 15 μg/liter as $CdCl_2$, and they consumed significantly more of the water that contained 12 μg/liter than they did of the control water. Surface waters rarely are found to contain more than 10 μg/liter (Durum *et al.*, 1971). Cadmium, not considered an essential element, is quite toxic and has been implicated in some human poisonings (Lieber and Welsch, 1954). However, interaction with other elements (Gunn and Gould, 1967; Hill *et al.*, 1963; Mason and Young, 1967; Fox and Reynolds, 1973) must be taken into account in considering its toxicity or in establishing limits for its concentrations in food, feed, and water. Cearly and Coleman (1973) demonstrated that when cadmium as cadmium sulfate was added to water at levels of 0.007, 0.09, and 0.83 mg/liter, concen-

TABLE 13 Recommended Limits of Concentration of Some Potentially Toxic Substances in Drinking Water for Livestock and Poultry[a]

Item	Safe Upper Limit of Concentration (mg/liter)
Arsenic	0.2
Barium	Not established[b]
Cadmium	0.05
Chromium	1.0
Cobalt	1.0
Copper	0.5
Cyanide	Not established[b]
Fluoride	2.0
Iron	Not established[b]
Lead	0.1
Manganese	Not established[b]
Mercury	0.010
Molybdenum	Not established[b]
Nickel	1.0
Nitrate - N	100.
Nitrite - N	10.
Salinity	See Table 10
Vanadium	0.1
Zinc	25.0

[a]For more detail on physiological effects of these elements, see Table 11 (drinking water) and Table 12 (feeds). The concentration values in Table 13 are generally far below the LD_{50} intakes of the various elements.

[b]No limit is given for a number of elements since experimental data available are not sufficient to make definite recommendations.

trations in the ash of naiad (*Najas quadulepensis*) weed ranged from 50 to 5,000 ppm, making the latter a potential source of cadmium in the food chain. Apparently through biological processes, muscle meat and milk are protected against accumulations of the element that would make them dangerous to man (Miller, 1971; Miller *et al.*, 1967; Doyle *et al.*, 1974).

Fluoride concentration in some groundwaters reaches about 15 mg/liter (Clarke and Clarke, 1967). According to Underwood's review (1971), chronic fluoride poisoning of livestock has been observed at this level. As little as 2 mg/liter apparently causes tooth mottling under some conditions without other harmful physiological effects. The element does not accumulate in soft tissues, and it is transferred only in small amounts to milk or eggs. In addition to Underwood's review, several other reports (Anon., 1966; Harris *et al.*, 1963; Saville, 1967; Schroeder *et al.*, 1968a; Shupe *et al.*, 1964) suggest that, except for tooth mottling, several mg/liter of fluoride in water causes no animal health problems.

Lead has long been known as a toxic element, but the level at which it becomes toxic has not been clearly established. This is still true even with the aid of some of the more recent reports (Aronson, 1971; Damron *et al.*, 1969; Donawick, 1966; Link and Pensinger, 1966; Harbourne *et al.*, 1968; Egan and O'Cuill, 1970; Hatch and Funnell, 1969). Link and Pensinger (1966) concluded that pigs can tolerate higher intakes of lead than cattle. They reported that 8 pigs consumed 11–66 mg/kg body wt of elemental lead in their feed without acute toxicosis. Hammond and Aronson (1964) have suggested 6–7 mg Pb/kg of body wt as a chronically toxic daily dose for cattle. They also reported that the element tends to accumulate in tissues and to be transferred to milk at levels that might harm consumers. Horses proved somewhat more susceptible than cattle. Damron *et al.* (1969) reported that broilers tolerated 100 ppm of dietary lead as acetate salt but had a decreased weight gain when fed 1,000 ppm. Lead pipe, once a common source of contamination, has through time been replaced by other materials. Natural waters usually contain less than 0.05 mg Pb/liter except possibly where galena minerals are in abundance (Durum *et al.*, 1971; Kopp and Kroner, 1962–1967). A recent thorough review of lead in our environment (NRC, 1972b) concludes that no evidence exists of any trend toward an increase of that element in natural waters or that it constitutes a health problem to fish.

Soluble inorganic *mercury* tends to absorb in sediment (Hem, 1970). The levels found in surface waters are, thus, usually very low (Durum *et al.*, 1971). However, biological methylation (Jensen and Jernelov, 1969) can slowly remove the element from sediment by converting it to methylmercury, which is stable, accumulates in body tissues (Gage, 1964; Miller *et al.*, 1961), and has a somewhat greater toxicity (Swensson *et al.*, 1959). Brain, liver, and kidney appear to accumulate the element more than other tissues (Aberg *et al.*, 1969; Dustman *et al.*, 1970), although this process is dependent upon its chemical form (Emerick and Holm, 1972). Inorganic mercury was poorly absorbed from the alimentary tract of goats and very little of that transferred to the milk (Howe *et al.*, 1972). Ansari *et al.* (1972) reported a 6.5 percent retention of orally administered radioactive methylmercury chloride in the liver, kidney, lung, and heart of calves as compared to a corresponding value of 0.42 percent for mercury chloride. Transfer of methylmercury to the fetus has been observed (Curley *et al.*, 1971), and the element has been found in eggs (Kiwimai *et al.*, 1969) and dry milk (Tanner *et al.*, 1972). Indirect poisoning by mercury compounds retained in animal tissues and subsequently consumed has been recorded for cases of man consuming pork (Curley *et al.*, 1971) or fish (Uchida *et al.*, 1961); ferrets consuming chickens (Hanke *et al.*, 1970); and rats fed organs from other rats (Ulfvarson, 1969). It appears, therefore, that while mercury in waters may not be a cause for concern over the health of livestock, the role of the latter as concentrators of the element for man's diet needs careful consideration.

Nitrate is a frequently encountered contaminant of water in rural areas and may occur in urban areas that have a high concentration of septic tanks. Analysis of more than 6,000 water samples for nitrate and nitrite in Missouri showed that 42 percent contained more than 5 ppm NO_3-N (Smith, 1965; Willrich and Smith, 1970). The nitrate content of well water correlated closely with hydrologic areas but had a limited relationship to soil types; no apparent correlation existed between the amount of fertilizer applied to the soil and nitrate accumulation in well waters. Coarse textured soils release more NO_3 to water than do fine textured soils (Maletic, 1973). Keller and Smith (1967) observed a band around feedlots where nitrogen had accumulated, a considerable part of which was present as nitrate. Nitrate in water correlated directly with the number of domestic animals and inversely with the depth of wells.

Nitrite was found in 1–2 percent of the wells in winter and 3–4 percent in summer. It was not correlated with livestock production nor with heavy fertilizer use except in sand point wells in river flood plains (Smith, 1965). Soils beneath feedlots contain nitrogen, and bacteria are present at depths of 15–30 ft in some types of soil. Intermediate products, such as ammonia and nitrite, tend to reach maximal amounts in the profile, whereas urea declines and nitrate, in the absence of denitrification, will accumulate. Below this depth, the soil is essentially sterile and no further change in water draining would then be present. The nitrate would ultimately be released at some point further downstream. Wells in limestone areas seem to be particularly high in nitrate (Smith, 1965).

Byers (1935) reported that 9 out of 18 samples of water from Nebraska and South Dakota had no detectable *selenium* present; the others, with one exception, ranged up to 1.2 mg/liter. To date, no documented case of selenium poisoning by water is known. Byers *et al.* (1938), however, discussed the possibility that water from a well and certain irrigation ditches in Colorado could contribute significantly to total intake of animals grazing seleniferous forage. Byers (1936) concluded that it was not likely that potable water would be found toxic due to its selenium content.

The role of selenium as a nutrient as well as a toxicant has recently been summarized (NRC, 1971c). Dietary levels of 0.02 ppm for ruminants and 0.03–0.05 ppm for poultry are suggested as essential for the prevention of a deficiency of the element. In addition, experimental evidence supports dietary levels of 0.1 ppm for livestock and 0.2 ppm for turkeys as safe for the prevention of the deficiency.

How likely are water supplies to reach the toxic levels of the substances that have been discussed? As shown in the section on nutrients in water for livestock and poultry (pp. 29–38), few waters supplied at mean concentrations provide as much as 10 percent of the daily nutrient needs. Only NaCl, magnesium, manganese, sulfur, and iodine at maximum concentration in surface waters of the United States exceeded 100 percent of the dietary requirement in the daily water intake of some species of livestock and poultry. It seems unlikely, therefore, that water toxic from excesses of nutrient elements will be a significant factor in the livestock industry.

However, Lunin (1970) emphasized that the toxic effects of the contaminants on animals should be differentiated from biological accumulations in the tissues that may subsequently be passed on to humans.

A number of the essential micronutrients do not increase with the age of the animal, despite exposures to large amounts of the elements in water. These are iron, manganese, zinc, copper, cobalt, and molybdenum. Vanadium and nickel also do not increase with age, except possibly in the lungs. Certain trace minerals such as cadmium, lead, titanium, and tin do accumulate with age (Schroeder *et al.*, 1963a,b).

Kirchgessner *et al.* (1967) reviewed the influence of trace and major element intakes of cows on the concentration of the elements in milk. While the intake of increased amounts of certain trace elements slightly influenced the concentrations of some of the trace elements, it was apparent that the milk should not provide a toxic level of minerals to humans.

Contamination from point sources should probably be our main concern for the foreseeable future. Pollutants described by Stokinger (1969) as being of concern include arsenic, fluorides, nitrates, asbestos, and hardness. Air-based contaminants that might reach water are beryllium and lead; food-based elements are cadmium, chromium, lead, and selenium.

The microcations applied to control plant disease and pests in field crops have been copper, arsenic, and mercury. Copper equivalent to about one-fourth of that added to the ocean each year through river water is applied to the soil. The use of arsenic is now declining. Mercury has been applied at a rate equivalent to about 1 percent of the annual loss to the ocean (Bowen, 1968), but its use is also declining.

Pollution from mining operations is generally somewhat localized, as is industrial pollution. Industrial pollution and the burning of coal contributes copper, zinc, and lead to the atmosphere (Warren *et al.*, 1969).

Various industries and mines contribute arsenic, lead, fluorine, and occasionally copper and molybdenum; tanneries use chromium. Approximately 2 pounds of lead per capita are added to gasoline each year; 1 pound of that amount is subsequently emitted in exhaust and contaminates roadsides (Schroeder, 1967). Chow and Earl (1970) presented data to show that the concentration of atmospheric lead was increasing at the rate of 5 percent per year in a metropolitan area. Mine slag from blast furnaces may contain very high levels of lead and is sometimes used as a traction device on roads. Rains subsequently may carry the lead into nearby streams.

The two main air contaminants resulting from burning oil and coal are carbon dioxide and sulfur dioxide. The amount of carbon dioxide in the atmosphere appears to have increased by 14 percent in the last 70 years, but as yet no measurable biological consequences have resulted (Bowen, 1968). Trace elements are present in the air and are washed from the atmosphere by rainfall (United States Public Health Service, 1964–1965).

Trace elements could be added to water from the distribution system. Asbestos and lead have long been recognized as toxicants. Glass, plastics, and stainless steel can also contribute several elements (Moore and Leddicotte, 1968).

Consideration of the implications of stocking many animals in confined systems and of recycling excrement adds a new dimension. The concept of using some "insurance levels of nutrients" and the use of salt to control feed intake will need to be evaluated further. The effect of mineral elements being placed into waste systems that will increase their concentrations, as well as their return to forage that will be used to feed a new generation of animals, will also need to be determined. For these reasons the study of elemental cycles is very important. In studying contamination, only the amount of a substance needs to be measured, whereas, in studying cycles, both the amount and the rate of change with respect to time should be measured (Allaway, 1968; Perry, 1968; Mertz, 1968; Bowen, 1968; Hodgson, 1969). Such information is needed to prevent the possibility of excessive levels of minerals accumulating in water and/or feed.

Toxic Algae

The poisoning of livestock by accumulations of certain of the blue-green algae in lake waters was recognized during the late 1800's (Francis, 1878). Cattle, sheep, and poultry apparently have been affected (Fitch *et al.*, 1934), and poisonings have been reported in many of our North Central states and in a number of other countries (Gorham, 1960, 1964).

Gorham (1964) lists six species, as follows, as potential causes of livestock poisonings: *Nodularia spumingena* Mert.; *Aphanizomenon flos-aquae* (L.) Ralfs; *Coelosphaerium kutzingianum* Nageli; *Gloeotrichia echinulata* (J. E. Smith) Richter; *Microcystis aeruginosa* Kutz. *emend.* Elenkin; and *Anabaena flos-aquae* (Lyngb.) de Breb. Of these, he reports the last two as most often involved in serious poisonings.

Bishop *et al.* (1959) reported the isolation from *Microcystis aeruginosa* NRC-1 of a toxic peptide containing D-serine, which they identified as the fast death factor (FDF) because it caused death quickly. Another factor, the slow death factor (SDF), is yet to be identified and causes a slower death. Predeath symptoms in livestock have not been carefully described. Postmortem examination is apparently of no help in diagnosis (Fitch *et al.*, 1934).

It has been observed (Shilo, 1967) that a sudden decomposition of

algal blooms, especially during periods of stagnation, often precedes
livestock poisonings. This suggests botulism, although it just as possibly indicates a release of toxins upon lysis of the cells. Identification
of any of the toxic blue-green algae species in suspect waters does no
more than suggest them as a cause of livestock deaths. In view of the
many unknowns related to toxic algal blooms, one can conclude that
water with heavy algal growth should best be avoided.

Radionuclides

Surface groundwaters acquire radioactivity from natural sources, from
fallout as a result of atmospheric nuclear detonations, from mining or
processing of uranium, or as the result of the use of isotopes in medicine, scientific research, or industry.

All radiation is regarded as harmful, and any unnecessary exposure
to it should be avoided. Experimental work on the biological half-lives
of radionuclides and their somatic and genetic effects on animals have
been briefly reviewed by McKee and Wolf (1963). Because the rate of
decay of a radionuclide is an unalterable property that cannot be
changed by known chemical or physical means, radioactive isotopes
must be disposed of by dilution or by storage and natural decay.

Based on the recommendations of the U.S. Federal Radiation Council
(1961), the U.S. Public Health Service has set drinking water standards
for radionuclides (U.S. Department of Health, Education, and Welfare,
1962a) at a level that the intake of radioactivity from these waters,
when added to that from all other sources, is not likely to be harmful
to man. Supplies containing radium-226 and strontium-90 are acceptable
provided that the concentrations of these radionuclides do not exceed
3 and 10 pCi/liter, respectively. In the known absence of strontium-90
and the alpha-emitting radionuclides, the water supply is considered
acceptable if the gross beta decay activity does not exceed 1,000 pCi/
liter. Gross beta-emitting concentrations in excess of this are grounds
for rejection of the supply, except when more complete analysis indicates that the radionuclide concentrations are not likely to cause exposures in excess of those adopted after recommendation by the U.S.
Federal Radiation Council (1961).

More recently, the National Academy of Sciences (1972) has reported
guidelines as follows for public water supplies, recognizing that these
should not be considered absolute maxima:

Strontium-90	2.5 pCi/liter
Radium-226	0.25 pCi/liter
Tritium	3,000 pCi/liter
All other radionuclides	1/150 of the limit for continuous occupational exposure set by the International Commission on Radiological Protection

Pesticides in Water for Livestock

Pesticides enter water from soil runoff, drift, rainfall, direct application, accidental spills, or faulty waste disposal techniques (Nicholson, 1970; Timmons et al., 1970). The use of pesticides in agriculture presents a potential hazard to livestock (An Der Lan, 1966). Compounds such as organophosphorus insecticides can be very dangerous. Pesticide classes that pose possible hazards are acaricides, fungicides, herbicides, insecticides, molluscides, and rodenticides (Papworth, 1967).

Acaricides recommended for use on crops and trees usually have low toxicity for livestock. Some, such as chlorobenzilate, have significant toxicity for mammals. With fungicides, the main hazard to livestock is not from the water route, but from their use as seed dressings for grain. The use of all organomercury fungicides is restricted by the Environmental Protection Agency (1972). Consequently, the possible hazard to livestock from these compounds has markedly decreased.

Schwartz et al. (1973) observed that 16 common pesticides had no effect on in vitro rumen microbe digestion of dry matter and cell wall

TABLE 14 Solubility, Toxicity, and Concentrations of Pesticides Observed in Water

Item	Solubility (μg/liter)[a]	Toxicity LD_{50}	Maximum Concentration (μg/liter)[b]
Aldrin	200	Rat: 80 mg/kg[c]	0.085
Dieldrin	250	Rat: 80 mg/kg[c]	0.407
Endrin	230	Rat: 9 mg/kg[c]	0.133
Heptachlor	negligible	Male rats: 60 mg/kg[d]	0.048
DDT	40	Cowbirds: 500 ppm in diet, 6–8 days[e]	0.316
DDE	–	Cowbirds: 1,500 ppm in diet, 19 days[e]	0.050
DDD	negligible	Cowbirds: 1,500 ppm in diet, 10.5 days[e]	0.840
2, 4-D	53,000	Rat: 375–805 mg/kg[f]	–

[a] Gunther et al., 1968.
[b] Lichtenburg et al., 1969.
[c] Keplinger and Deichmann, 1967.
[d] Eisler, 1970.
[e] Stickel et al., 1970.
[f] Way, 1969.

constituents when present at concentrations below 100 ppm in the medium. Insecticides of vegetable origin, such as pyrethrins and rotenones, are generally believed to be practically nontoxic to livestock. The toxicities of DDT, DDD, dilan, methoxychlor, and perthane are low for mammals, while other insecticides may be more toxic (Papworth, 1967; Radeleff, 1970). The organophosphorus insecticides are potentially the most hazardous due to their cholinesterase inhibition. Mipafax has been shown to cause other pathological changes (Barnes and Denz, 1953). The liquid organophosphorus insecticides are absorbed by all routes and their lethal doses are quite low (Radeleff, 1970).

Some solubility values, toxicity (LD_{50}, or dose sufficient to kill half the animals), and observed maximum concentrations of several common pesticides in water are shown in Table 14. The maximum concentrations in surface water observed for any of the pesticides was considerably below the LD_{50} doses for the animals, which only indicates that their health is not threatened. Their meat and edible products may be contaminated. Although many pesticides are readily broken down and eliminated by livestock with no subsequent toxicological effect, the inherent problems associated with pesticide use include their secretion in milk, as well as accumulation in edible tissues (Crosby, *et al.,* 1967; Waldron *et al.,* 1968). Interactions between insecticides and drugs are also a possibility, especially in animal feeds (Conney and Hitchings, 1969).

The subject of toxic levels of pesticides and herbicides in water for livestock and other agricultural uses was reviewed by Edwards (1970), Little (1970), and the National Academy of Sciences–National Academy of Engineering (1972).

SUMMARY

The geochemical cycle of water has been reviewed in regard to the effects of evaporation, precipitation, ground, and oceans on the natural constituents of water. Fifty-eight mineral constituents have been classified into major, secondary, minor, and trace categories. Ten elements make up approximately 99 percent of the dissolved minerals. Hydrogen-ion activity (pH), hardness, dissolved salts, color, turbidity, biological and chemical oxygen demand, taste, odor, and temperature have been pointed out as general physical and chemical properties of water. Water in the United States has been classified into four chemical types: calcium–magnesium, carbonate–bicarbonate; calcium–magnesium, sulfate–chloride; sodium–potassium, carbonate–bicarbonate; and sodium–potassium, sulfate–chloride. Eighty-seven percent of the water in the United States is dominated by calcium and magnesium, leaving only 13 percent belonging to the sodium–potassium type. The former usually occur with carbonate and bicarbonate and to a smaller extent with chloride and sulfate. Generally, sodium and potassium are combined with sulfates and chlorides and to a lesser extent with carbonates and bicarbonates.

The water requirements of beef and dairy cattle, sheep, swine, horses, and poultry have been discussed in regard to effect of ambient temperature and some physiological factors. Beef cattle weighing 450 kg may drink approximately 27, 39, and 63 liters of water per day at 4, 21, and 32 °C, respectively. Yearling feedlot cattle may consume 50 percent more water in summer than in winter. Dairy cows producing 40 kg of milk per day may drink up to 110 kg of water when fed dry feeds and may suffer more quickly from a lack of water than from a shortage of any other nutrient. Generally, water consumption by sheep amounts to two times the weight of dry matter intake; but ambient temperature, activity, type of pasture, and other factors may affect this value. Ewes in winter on dry feed require 4 liters per head daily prior to lambing and 6 or more liters per day when nursing lambs. Swine generally require 2–2.5 liters of water per kg of dry feed, but may drink 4–4.5 liters in hot weather. Horses as a rule need 2–3 liters of water per kg dry feed, but composition of feed, ambient temperature, and activity may greatly influence the requirements. Laying hens are very dependent on an adequate supply of water, and young broilers rapidly increase their water intake with growth.

Nutrient minerals in water were discussed on the basis of the percentage of the National Research Council requirements for livestock species and poultry that are present in average intakes at mean and maximum concentrations. Salt as sodium chloride at average concentrations could supply 6–34 percent of the requirements of various species of livestock and poultry. If maximum concentrations of sodium chloride are present, approximately 8–54 times the daily requirements of the various species would be present. The consequences of such excesses are discussed under salinity. Calcium at mean concentrations would supply in the average daily intake of water 5–28 percent of the requirements of sheep, cattle, and horses; swine and poultry, 3 percent or less. At maximum concentrations, sufficient calcium is present to supply 15–86 percent of the requirements of beef and dairy cattle, sheep, and horses but only 1–10 percent of those of swine and poultry. At average concentrations of phosphorus in water, less than 1 percent of the daily requirements of livestock would be supplied. At maximum concentrations only 1–3 percent of the daily requirements for the various species would be present in the water consumed. Magnesium at the mean concentration would provide 4–11 percent of the needs of beef and dairy cattle, sheep, swine, horses, chickens, and turkeys. At maximum levels, 42–110 percent of the magnesium requirements of the various species would be provided. Less than 1 percent of the potassium requirements of beef and dairy cattle, swine, and horses would be present at mean concentrations and 24–53 percent at maximum concentrations. Sulfur at mean concentrations in drinking water would supply 10–38 percent of the requirements of cattle, sheep, and horses and

2-11 times the daily dietary requirements at maximum levels. Approximately 1 percent of the daily requirements of iron for beef and dairy cattle, swine, and poultry are present at mean concentrations compared to approximately 12-60 percent at maximum concentrations.

Mean concentrations of zinc in drinking water would provide 1-2 percent of the requirements of beef and dairy cattle and sheep and less for swine and poultry. At maximum concentrations, 12-51 percent of the requirements of beef and dairy cattle and 3-6 percent of the requirements of sheep, swine, and poultry would be present. Copper at average concentrations could meet 1-2 percent of the daily requirements of the six species reported, while at maximum concentrations 9-33 percent would be supplied daily in normal water consumption. At the mean level of cobalt in drinking water, 3-12 percent of the daily requirements of beef and dairy cattle, sheep, and horses would be supplied; at maximum concentrations, 15-63 percent. Manganese at average concentrations would supply approximately 3-6 percent of the daily dietary requirements of beef and dairy cattle and less than 1 percent of those of swine and poultry; at maximum levels, 3-6 times the requirements of beef and dairy cattle, 18-39 percent of those of swine, and 11 percent of those of poultry. Selenium at mean concentrations provides approximately 1 percent of the dietary requirements of beef and dairy cattle, sheep, swine, and poultry. Maximum levels would supply approximately 1-10 percent of the requirements for these species. Due to lack of data on iodine concentration in water in the United States, only Florida values were used for illustrative purposes. Water in Florida, generally, has a very significant amount of iodine present for meeting the needs of livestock and poultry.

Potentially toxic substances in drinking water of livestock and poultry were discussed in regard to their effect on growth, reproduction, longevity, and build-up in edible tissues and products when data were available. Since extensive studies on the effects of most toxicants on livestock and poultry have not been conducted, data on various experimental species were also presented when available. A number of elements, such as iron, cobalt, copper, chromium, manganese, molybdenum, iodide, and zinc, seldom cause problems when in the drinking water of livestock and poultry, either from effects on production or build-up in their tissues. Experimental levels at which these elements have been observed to cause problems with various species of animals were summarized. Arsenic, nitrate, and selenium were discussed rather extensively due to more data being available, as well as to their reputation as toxicants. However, these three substances in drinking water have seldom been demonstrated to cause harm to livestock and poultry. Other elements, such as lead, mercury, and cadmium, are more hazardous to livestock and poultry, especially due to build-up in their tissues and products at levels undesirable to persons that consume them.

Effects of various salts at high concentrations in water were discussed in regard to six species of farm animals. Water that contains less than 1,000 mg/liter of total dissolved salts should present no serious problems to any class of livestock or poultry. Water that contains 1,000–2,999 mg/liter should be satisfactory for all species of livestock and poultry in regard to performance, though some mild and temporary diarrhea may occur. When the water contains 3,000–4,999 mg/liter, it is of poor quality for poultry and at the higher levels may cause increased mortality and decreased growth. However, livestock should find this range of salinity satisfactory, especially when they become accustomed to it. Water in the range of 5,000–6,999 mg/liter can be used with reasonable safety for beef and dairy cattle, sheep, swine, and horses, although it is best to avoid higher levels for pregnant and lactating animals. Salinity in this range is not acceptable to poultry. In the range of 7,000–10,000 mg/liter of saline salts, the waters are unfit for poultry and probably for swine. They are a source of risk for pregnant and lactating cows, sheep, and horses, as well as for the young of these species and those subjected to heat stress. Waters that contain more than 10,000 mg/liter of saline salts involve sufficient risk that they probably should not be used.

Toxic blue-green algae were pointed out as a worldwide problem in drinking water for livestock. To date only one toxin has been reported as isolated and identified. It is a cyclic polypeptide containing 10 amino acid residues, one of which is the unnatural amino acid D-serine. The sudden decomposition of algal blooms often precedes mass mortality of fish and these decompositions have been associated with livestock poisonings. Predeath symptoms due to algal poison have not been well observed and postmortem examination is apparently of no help in diagnosis. In view of the many unknowns relating to toxic algae blooms, the use of drinking water with heavy growths should best be avoided.

Radionuclides occur in water from both natural and human sources. In general, the radioactivity of drinking water for livestock and poultry should be of no greater level than that recommended for human consumption by the U.S. Public Health Service.

Limited information on the effects of pesticides in water on economic animals and their products was presented and their potential hazards pointed out. Recommendations are given in Table 13 on limits of concentration of some potential toxic substances in drinking water for livestock and poultry.

REFERENCES

Aberg, B., L. Ekman, R. Falk, U. Greitz, G. Persson, and J. O. Snihs. 1969 Metabolism of methyl mercury (^{203}Hg) compounds in man. Excretion and distribution. Arch. Environ. Health 19:478-84.

Adams, A. W., R. J. Emerick, and C. W. Carlson. 1966. Effects of nitrate and nitrite in the drinking water on chicks, poults and laying hens. Poult. Sci. 45:1215-22.

Adolf, E. F. 1933. The metabolism and distribution of water in body and tissues. Physiol. Rev. 13:336-71.

Allaway, W. H. 1968. Control of environmental levels of selenium, pp. 181-206. In D. D. Hemphill, ed. Proc. 2nd Annu. Conf. Trace Subst. Environ. Health, July 16-18, 1968. Columbia: University of Missouri.

Allcroft, R. 1951. Lead poisoning in cattle and sheep. Vet. Rec. 63:583.

Allcroft, R., and K. L. Blaxter. 1950. The toxicity of lead to cattle and sheep and an evaluation of the lead hazard under farm conditions. J. Comp. Pathol. Ther. 60:209-18.

Allison, I. S. 1930. The problem of saline drinking waters. Science 71:559-60.

Alsmeyer, R. H., T. J., Cunha, and H. D. Wallace. 1955. Preliminary observations on the effect of source of water on rate of gain of growing fattening pigs. Fla. Agric. Exp. Stn. Anim. Husb. Mimeo Ser. No. 55-5.

An Der Lan, H. 1966. The present situation of toxicology in the field of crop production. F. A. Gunther, ed. Residue Rev. 15:31-43.

Andrews, E. D. 1965. Colbalt poisoning in sheep. N. Z. Vet. J. 13:101-3.

Angino, E. E., O. K. Galle, and T. C. Waugh. 1969. Fe, Mn, Ni, Co, Sr, Li, Zn and SiO$_2$ in streams of the Lower Kansas River Basin. Water Resour. Res. 5 (3):698-705.

Anon. 1950. Water for agricultural purposes in Western Australia. J. Agric. West. Aust. 27:156-60.

Anon. 1959. Salinity and livestock water quality. S. D. State Coll. Bull. 481. 12 pp.

Anon. 1966. Carbohydrate metabolism of rats consuming 450 p.p.m. fluoride. Nutr. Rev. 24:346–47.

Ansari, M. S., W. J. Miller, M. W. Neathery, R. P. Gentry, P. E. Stake, and J. W. Lassiter. 1972. Metabolism of two forms of ^{203}Hg by calves. J. Anim. Sci. 35:184. (abstr.)

Aronson, A. L. 1971. Lead poisoning in cattle and horses following long-term exposure to lead. J. Am. Vet. Med. Assoc. 158:1870. (abstr.)

Arrington, L. R. 1973. Personal communication.

Arrington, L. R., R. N. Taylor, C. B. Ammerman, and R. L. Shirley. 1965. Effects of excess dietary iodine upon rabbits, hamsters, rats and swine. J. Nutr. 87:394–98.

Arrington, L. R., R. A. Santa Cruz, R. H. Harms, and H. R. Wilson. 1967. Effects of excess dietary iodine upon pullets and laying hens. J. Nutr. 92:325–30.

Arthur, D., I. Motzok, and H. D. Branion. 1958. Interaction of dietary copper and molybdenum in rations fed to poultry. Poult. Sci. 37:1181. (abstr.)

Asplund, J. M., and W. H. Pfander. 1972. Effects of water restriction on nutrient digestibility in sheep receiving fixed water:feed ratios. J. Anim. Sci. 35:1271–4.

Back, W., and B. B. Hanshaw. 1965. Advances in Hydroscience, pp. 50–109. Vol. 2, Ven T. Chow, ed. New York: Academic Press.

Balch, C. C., D. A. Balch, V. W. Johnson, and J. Turner. 1953. Factors affecting the utilization of food by dairy cows. VII. The effect of limited water intake on the digestibility and rate of passage of hay. Br. J. Nutr. 7:212–24.

Ballantyne, E. E. 1957. Drinking waters toxic for livestock. Can. J. Comp. Med. Vet. Sci. 21:254–57.

Barnes, J. M., and F. A. Denz. 1953. Experimental demyelination with organophosphorus compounds. J. Pathol. Bacteriol. 65:597–605.

Becker, D. E., and S. E. Smith. 1951. The level of cobalt tolerance in yearling sheep. J. Anim. Sci. 10:266–71.

Berg, L. R., G. E. Bearse, and L. H. Merrill. 1963. Vanadium toxicity in laying hens. Poult. Sci. 42:1407–11.

Berg, R. T., and J. P. Bowland. 1960. Salt water tolerance of growing-finishing swine. 39th Annu. Feeder's Day Bull. Edmonton, Canada: University of Alberta.

Bishop, C. T., E. F. L. J. Anet, and P. R. Gorham. 1959. Isolation and identification of the fast-death factor in *Microcystis aeruginosa*. Can. J. Biochem. Physiol. 37:453–71.

Blaxter, K. L. 1950. Lead as a nutritional hazard to farm livestock II. Factors influencing the distribution of lead in the tissues. J. Comp. Pathol. Ther. 60:177–89.

Bocconi, G. and C. Bonessa. 1953. Histological changes in the thyroid after administration of water containing arsenic and iron. Arch. Sci. Med. 95:421–28 (C.A. 47:9496d).

Boswell, M. C., and J. V. Dickson. 1918. The absorption of arsenious acid by ferric hydroxide. J. Am. Chem. Soc. 40:1793–1801.

Bowen, H. J. M. 1968. Elementary cycles and pollution, pp. 171–79. In D. D. Hemphill, ed. Proc. 2nd Annu. Conf. Trace Subst. Environ. Health, July 16–18, 1968. Columbia: University of Missouri.

Bradley, W. B., H. F. Eppson, and O. A. Beath. 1940. Livestock poisoning by oat hay and other plants containing nitrate. Wyo. Agric. Exp. Stn. Bull. 241.

Brink, M. F., D. E. Becker, S. W. Terrill, and A. H. Jensen. 1959. Zinc toxicity in the weanling pig. J. Anim. Sci. 18:836–42.

Buchman, D. T., R. L. Shirley, and G. B. Killinger. 1968. Nitrate, ammonia and methemoglobin in sheep when fed millet containing different levels of molybdenum and copper. Soil Crop Sci. Soc. Fla. Proc. 28:209–15.

Bucy, L. L., U. S. Garrigus, R. H. Forbes, H. W. Norton, and W. W. Moore. 1955. Toxicity of some arsenicals fed to growing-fattening lambs. J. Anim. Sci. 14:435–45.

Burden, E. H. J. W. 1968. The toxicology of nitrates and nitrites with particular reference to the potability of water supplies. Analyst 86:429–33.

Byerrum, R. U., R. A. Anwar, and C. A. Hoppert. 1960. Toxicity of small amounts Cd and Cr in water. Tech. eau 14: No. 165, 24 (C.A. 54:25320c).

Byers, H. G. 1935. Selenium occurrence in certain soils in the United States with a discussion of related topics. USDA Tech. Bull. 482. Washington, D.C. 47 pp.

Byers, H. G. 1936. Selenium occurrence in certain soils in the United States with a discussion of related topics. USDA Tech. Bull. 530. Washington, D.C. 78 pp.

Byers, H. G., J. T. Miller, K. T. Williams, and H. W. Lakin. 1938. Selenium occurrence in certain soils in the United States with a discussion of related topics. USDA Tech. Bull. 601. Washington, D.C. 74 pp.

Byron, W. R., G. W. Bierbower, J. B. Brouwer, and W. H. Hansen. 1965. Pathological changes in rats and dogs from 2-year feeding of sodium arsenite or sodium arsenate. Fed. Proc. 393. (abstr.)

Cearly, J. E., and R. L. Coleman. 1973. Cadmium toxicity and accumulation in Southern Naiad. Bull. Environ. Contam. Toxicol. 9:100.

Chapman, H. L., Jr., S. L. Nelson, R. W. Kidder, W. L. Sippel, and C. W. Kidder. 1962. Toxicity of cupric sulfate for beef cattle. J. Anim. Sci. 21:960–62.

Chicco, C. F., C. B. Ammerman, P. A. van Walleghem, P. W. Waldroup, and R. H. Harms. 1967. Effects of varying dietary ratios of magnesium, calcium and phosphorus in growing chicks. Poult. Sci. 46:368–73.

Chow, T. J., and J. L. Earl. 1970. Lead aerosols in the atmosphere: Increasing concentrations. Science 169:577–80.

Clarke, E. G. C., and M. L. Clarke, eds. 1967. Garner's Veterinary Toxicology. 3rd ed. London: Bailliere, Tindall and Cassell, Ltd.

Clough, G. W. 1934. Poisoning of animals by cyanides present in some industrial effluents. Vet. Rec. 13:538.

Conney, A. H., and G. H. Hitchings. 1969. Combinations of drugs in animals feeds, pp. 180–192. In: The Use of Drugs in Animal Feeds. Symposium. NAS Publ. No. 1679. Washington, D.C.

Coup, M. R., and A. G. Campbell. 1964. The effect of excessive iron intake upon the health and production of dairy cows. N. Z. J. Agric. Res. 7:624–38.

Crawford, R. F., W. K. Kennedy, and K. L. Davison. 1966. Factors influencing the toxicity of forages that contain nitrate when fed to cattle. Cornell Vet. 56:3–17.

Crosby, D. G., T. E. Archer, and R. C. Laben. 1967. DDT contamination in milk following a single feeding exposure. J. Dairy Sci. 50:40–42.

Curley, A., V. A. Sedlak, E. F. Girling, R. E. Hawk, W. F. Barthel, P. E. Pierce, and W. H. Likosky. 1971. Organic mercury identified as the cause of poisoning in humans and dogs. Science 172:65–67.

Damron, B. L., C. F. Simpson, and R. H. Harms. 1969. The effects of feeding various levels of lead on the performance of broilers. Poult. Sci. 48:1507–9.

Dantzman, C. L., and H. L. Breland. 1969. Chemical status of some water sources in south central Florida. Soil Crop Sci. Soc. Fla. Proc. 29:18–28.

Davies, R. E., B. L. Reid, A. A. Kurnick, and J. R. Couch. 1960. The effect of sulfate on molybdenum toxicity in the chick. J. Nutr. 70:193–98.

Davis, G. H., J. H. Green, F. H. Olmsted, and D. W. Brown. 1959. Groundwater conditions and storage capacity, San Joaquin Valley. USGS Water-Supply Paper 1469. 168 pp.

Davis, G. K. 1950. A Symposium on Copper Metabolism, Baltimore: The Johns Hopkins Press. 443 pp.

Davis, S. N., and R. J. M. DeWiest. 1966. Hydrogeology. New York: John Wiley & Sons. 112 pp.

Davison, K. L., W. Hansel, L. Krook, K. McEntee, and M. J. Wright. 1964. Nitrate toxicity in dairy heifers. I. Effects on reproduction, growth, lactation and vitamin A nutrition. J. Dairy Sci. 47:1065–73.

Davison, K. L., K. McEntee, and M. J. Wright. 1965. Responses in pregnant ewes fed forages containing various levels of nitrate. J. Dairy Sci. 48:968–77.

Decker, L. E., R. U. Byerrum, C. F. Decker, C. A. Hoppert, and R. F. Langham. 1958. Chronic toxicity studies. I. Cadmium administered in drinking water to rats. AMA Arch. Ind. Health 18:228–31.

Deobold, H. J., and C. A. Elvehjem. 1935. The effect of feeding high amounts of soluble iron and aluminum salts. Am. J. Physiol. 111:118–23.

Diven, R. H., R. E. Reed, and W. J. Pistor. 1964. The physiology of nitrite poisoning in sheep. Ann. N.Y. Acad. Sci. 111:638–43.

Doberenz, A. R., A. A. Kurnick, B. J. Hulett, and B. L. Reid. 1965. Bromide and fluoride toxicities in the chick. Poult. Sci. 44:1500–4.

Doherty, P. C., R. M. Barlow, and K. W. Angus. 1969. Spongy changes in the brains of sheep poisoned by excess dietary copper. Res. Vet. Sci. 10:303–4.

Donawick, W. J. 1966. Chronic lead poisoning in a cow. J. Am. Vet. Med. Assoc. 148:655–61.

Doyle, J. J., and W. H. Pfander. 1974. Acceptability by lambs of cadmium in feed and water. Nutr. Rep. Int. 9:273–76.

Doyle, J. J., W. H. Pfander, S. E. Grebing, and J. O. Pierce II. 1974. Effect of dietary cadmium on growth, cadmium absorption and cadmium tissue levels in growing lambs. J. Nutr. 104:160–66.

Druckrey, H., D. Steinhoff, H. Beuthner, H. Schneider, and P. Klaerner. 1963. Screening of nitrite for chronic toxicity in rats. Arzneim.-Forsch. 13:320–23 (C.A. 59:6885h).

Durfor, C. N., and E. Becker. 1964. Public water supplies of the 100 largest cities of the United States, 1962. USGS Water-Supply Paper 1812, Washington, D.C.: U.S. Government Printing Office. 364 pp.

Durum, W. H., and J. Haffty. 1961. Occurrence of minor elements in water. USGS Geol. Surv. Circ. 445, Washington, D.C. 11 pp.

Durum, W. H., J. D. Hem, and S. G. Heidel, 1971. Reconnaissance of selected minor elements in surface waters of the United States, October 1970. USGS Geol. Surv. Circ. 643, Washington, D.C. 40 pp.

Dustman, E. H., L. F. Stickel, and J. B. Elder. 1970. Mercury in wild animals, Lake St. Clair. Proc. Int. Conf. Environ. Mercury Contam., Ann Arbor, Mich., Sept. 30–Oct. 2.

Edwards, Clive A. 1970. Persistent Pesticides in the Environment. Cleveland, Ohio: Chemical Rubber Co. Press. 78 pp.

Egan, D. A., and T. O'Cuill. 1970. Cumulative lead poisoning in horses in a mining area contaminated with galena. Vet. Rec. 86:736-37.

Eisler, Milton. 1970. Inter-American Conferences on Toxicology and Occupational Medicine, pp. 231-34. W. E. Deichman, J. L. Radomski, and R. A. Penalver, eds. Miami, Fla.: Halos and Assoc., Inc.

Embry, L. B., M. A. Hoelscher, R. C. Wahlstrom, C. W. Carlson, L. M. Krista, W. R. Brosz, G. F. Gastler, and O. E. Olson. 1959. Salinity and livestock water quality. S. D. Agric. Exp. Stn. Bull. 481. 12 pp.

Emerick, R. J., L. B. Embry, and R. W. Seerley. 1965. Rate of formation and reduction of nitrite-induced methemoglobin in vitro and in vivo as influenced by diet of sheep and age of swine. J. Anim. Sci. 25:221-30.

Emerick, R. J. and A. M. Holm. 1972. Toxicity and tissue distribution of mercury in rats fed various mercurial compounds. Nutr. Rep. Int. 6:125-31.

Environmental Protection Agency. 1972. Mercurial pesticides, man and the environment. Office of Pesticides Programs. Washington, D.C. 118 pp.

Fain, P., J. Dennis, and F. G. Harbaugh. 1952. Influence of added Mn in feed on various mineral components of cattle blood. Am. J. Vet. Res. 13:348-50.

Feth, J. H., et al. 1965. Preliminary map of the conterminous United States showing depth to and quality of shallowest ground water containing more than 1,000 parts per million dissolved solids. USGS Hydrol. Invest. Atlas HA-199, Washington, D.C.

Fitch, C. P., L. M. Bishop, W. L. Boyd, R. A. Gortner, C. F. Rogers, and J. E. Tilden. 1934. "Water bloom" as a cause of poisoning in domestic animals. Cornell Vet. 24:30-39.

Fonnesbeck, Paul V. 1968. Consumption and excretion of water by horses receiving all hay and hay-grain diets. J. Anim. Sci. 27:1350-56.

Fox, M. R. Spivey, and H. Reynolds. 1973. Trace minerals in foods: Nutritional toxicological problems. Feedstuffs 45(No. 13):30 and 51.

Francis, George. 1878. Poisonous Australian lake. Nature 18:11-12.

Frens, A. M. 1946. Salt drinking water for cows. Tijdschr. Diergeneeskd. 71 (No. 1):6-11 (C.A. 44:5972-3, 1950).

Frost, D. V. 1967. Arsenicals in biology—retrospect and prospect. Fed. Proc. 26:194-208.

Gage, J. C. 1964. Distribution and excretion of methyl and phenyl mercury salts. Br. J. Ind. Med. 21:197-202.

Garrels, R. M., and C. L. Christ. 1965. Solutions, minerals, and equilibria. New York: Harper & Row. 94 pp.

Gastler, G. F., and O. E. Olson. 1957. Dugout water quality. S. D. Farm Home Res. 8:20-23.

Glendening, B. L., W. G. Schrenck, and D. B. Parrish. 1956. Effects of rubidium in purified diet fed rats. J. Nutr. 60:563-79.

Goldschmidt, V. M. 1933. Grundlagen der quantitativen Geochemie. Fortschr. Miner. Krist. Petrog. 17:112-56.

Goodrich, R. D., R. J. Emerick, and L. B. Embry. 1964. Effect of sodium nitrate on the vitamin A nutrition of sheep. J. Anim. Sci. 23:100-4.

Gorham, P. R. 1960. Toxic waterblooms of blue-green algae. Can. Vet. J. 1:235-45.

Gorham, P. R. 1964. Toxic algae, pp. 307-36. In D. F. Jackson, ed. Algae and Man. New York: Plenum Press.

Greeson, P. E. 1970. Biological factor in the chemistry of mercury, pp. 32-34. In:

Mercury in the Environment. USGS Prof. Pap. No. 713, Washington, D.C.: U.S. Government Printing Office.

Gross, W. G., and V. G. Heller. 1946. Chromates in animal nutrition. J. Ind. Hyg. Toxicol. 28:52–56.

Grummer, R. H., O. G. Bentley, P. H. Phillips, and G. Bohstedt. 1950. The role of Mn in growth, reproduction and lactation of swine. J. Anim. Sci. 9:170–75.

Grushko, Y. M., V. A. Donskov, and V. S. Kolesnik. 1951. Toxicity of drinking water containing cadmium salts. Gig. Sanit. 9:22–26 (C.A. 46:4119d, 1952).

Gunn, S. A., and T. C. Gould. 1967. Specificity of response in relation to cadmium, zinc and selenium, pp. 395–413. In O. H. Muth, J. E. Oldfield, and P. H. Weswig, eds. Symposium: Selenium in Biomedicine. Westport, Conn.: AVI Publishing Co., Inc.

Gunther, F. A., W. E. Westlake, and P. S. Jaglan. 1968. Residues of pesticides and other foreign chemicals in foods and feeds. Residue Rev. 20:1–148.

Hale, W. H. 1973. Personal Communication.

Hammond, P. B., and A. L. Aronson. 1964. Lead poisoning in cattle and horses in the vicinity of a smelter. Ann. N.Y. Acad. Sci. 111:595–611.

Hanke, E., K. Erne, H. Wanntorp, and K. Borg. 1970. Poisoning in ferrets by tissues of alkyl mercury-fed chickens. Acta. Vet. Scand. 11:268–82.

Harbin, R., F. G. Harbaugh, K. L. Neeley, and N. C. Fine. 1958. Effect of natural combinations of ambient temperature and relative humidity on the water intake of lactating and non-lactating dairy cows. J. Dairy Sci. 41:1621–27.

Harbourne, J. F., C. T. McCrea, and J. Watkinson. 1968. An unusual outbreak of lead poisoning in calves. Vet. Rec. 83:515–17.

Harris, L. E., R. J. Raleigh, M. A. Madsen, J. L. Shupe, J. E. Butcher, and D. A. Greenwood. 1963. Effect of various levels of fluorine, stilbestrol and oxytetracycline in the fattening ration of lambs. J. Anim. Sci. 22:51–55.

Hartman, R. H., G. Matrone, and G. H. Wise. 1955. Effect of high dietary Mn on hemoglobin formation. J. Nutr. 57:429–39.

Hatch, R. C., and H. S. Funnell. 1969. Lead levels in tissues and stomach contents of poisoned cattle: A fifteen-year summary. Can. Vet. J. 10:258–62.

Hathcock, J. N., C. H. Hill, and G. Matrone. 1964. Vanadium toxicity and distribution in chicks and rats. J. Nutr. 82:106–10.

Heller, V. G. 1932. Saline and alkaline drinking waters. J. Nutr. 5:421–29.

Heller, V. G. 1933. The effect of saline and alkaline waters on domestic animals. Okla. Agric. Exp. Stn. Bull. 217. 23 pp.

Heller, V. G., and C. H. Larwood. 1930. Saline drinking water. Science 71:223–24.

Hem, J. D. 1970. Chemical behavior of mercury in aqueous media, pp. 19–24. In: Mercury in the Environment. USGS Prof. Pap. No. 713. Washington, D.C.: U.S. Government Printing Office.

Hemingway, R. G., and A. MacPherson. 1967. Storage of Cu by blackface lambs fattened indoors on diets based on cereals. Proc. 45th Meet. Br. Soc. Anim. Prod., Harrogate. Anim. Prod. 9:282.

Hemphill, F. E., M. L. Kaeberle, and W. B. Buck. 1971. Lead suppression of mouse resistance to *Salmonella typhimurium*. Science 171:1031–32.

Hibbard, P. L. 1934. The significance of mineral matter in water. J. Am. Water Works Assoc. 26:884–90.

Hill, C. H., G. Matrone, W. L. Payne, and C. W. Barber. 1963. *In vivo* interactions of cadmium with copper, zinc and iron. J. Nutr. 80:227–35.

Hill, C. H., B. Starcher, and G. Matrone. 1964. Mercury and silver interrelationships with copper. J. Nutr. 83:107-10.

Hinshaw, W. R., and W. E. Lloyd. 1931. Studies on the use of copper sulfate in turkeys. Poult. Sci. 10:392-23.

Hobbs, C. S., and G. M. Merriman. 1962. Fluorosis in beef cattle. Tenn. Agric. Exp. Stn. Res. Bull. 351. 183 pp.

Hodgson, J. F. 1969. Chemistry of trace elements in soils with reference to trace element concentration in plants, pp. 45-58. In D. D. Hemphill, ed. Proc. 3rd Annu. Conf. Trace Subst. Environ. Health, July 16-18, 1968. Columbia: University of Missouri.

Hoffman, M. P., and H. L. Self. 1972. Factors affecting water consumption by feedlot cattle. J. Anim. Sci. 35:871-7.

Holst, W. O., L. M. Flynn, G. B. Garner, and W. H. Pfander. 1961. Dietary nitrate vs. sheep performance. J. Anim. Sci. 20:936. (abstr. 145)

Howe, Sister M., J. McGee, and F. W. Lengemann. 1972. Transfer of inorganic mercury to milk of goats. Nature 237:316.

Hutagalung, R. I., C. H. Chaney, R. D. Wood, and D. G. Waddill. 1968. Effects of nitrates and nitrites in feed on utilization of carotene in swine. J. Anim. Sci. 27:79-82.

Hutchinson, G. E. 1957. A. Treatise on Limnology. New York: John Wiley & Sons. 1016 pp.

Hutchinson, G. L., and F. G. Viets, Jr. 1969. Nitrogen enrichment of surface water by absorption of ammonia volatilized from cattle feed lots. Science 166:514-15.

Iacovoni, P. F. D., R. Cassoni, V. Lucisano, and G. Gambelli. 1968. Interference of reserpine with some sodium nitrate effects on rabbits. Gazz. Int. Med. Chir. 73(24) (Pt. 3):5964-68 (C.A. 73:43656, 1970).

James, E. C., Jr., and R. S. Wheeler. 1949. Relation of dietary protein content to water intake, water elimination and amount of cloacal excreta produced by growing chicks. Poult. Sci. 28:465-67.

Jensen, M. C., G. C. Lewis, and G. O. Baker. 1951. Characteristics of irrigation waters in Idaho. Agric. Exp. Stn. Res. Bull. No. 19. Moscow: University of Idaho.

Jensen, S., and A. Jernelov. 1969. Biological methylation of mercury in aquatic organisms. Nature 223:753-54.

Johnson, D., Jr., A. L. Mehring, Jr., F. X. Savino, and H. W. Titus. 1962. The tolerance of growing chickens for dietary zinc. Poult. Sci. 41:311-17.

Jones, I. R., P. H. Weswig, J. F. Bone, M. A. Peters, and S. O. Alpan. 1966. Effect of high-nitrate consumption on lactation and vitamin A nutrition of dairy cows. J. Dairy Sci. 49: 491-99.

Kare, M. R., and J. Biely. 1948. The toxicity of sodium chloride and its relation to water intake in baby chicks. Poult. Sci. 27:751-58.

Keener, H. A., G. P. Percival, K. S. Morrow, and G. H. Ellis. 1949. Cobalt tolerance in young dairy cattle. J. Dairy Sci. 32:527-33.

Keller, W. D., and G. E. Smith. 1967. Ground water contamination by dissolved nitrate. Spec. Pap. No. 90, pp. 47-59. The Geological Society of America.

Kennedy, V. C. 1965. Mineralogy and cation-exchange capacity of sediments from selected streams. USGS Prof. Pap. 433-D. Washington, D.C. 28 pp.

Keplinger, M. L., and W. B. Deichmann. 1967. Acute toxicity of combination of pesticides. Toxicol. Appl. Pharmacol. 10:58-95.

Kienholz, E. W., D. K. Schisler, C. F. Nockels, and R. E. Moreng. 1966. Sodium and potassium nitrates in drinking water for turkeys. Poult. Sci. 45:1097. (abstr.)

Kirchgessner, M., H. Friescke, and G. Koch. 1967. Major and trace elements in milk, Chapter 6. Nutrition and Composition of Milk. London: Crosby Lockwood and Son, Ltd.

Kiwimai, A., A. Swensson, U. Ulfarson, and G. Westoo. 1969. Methylmercury compounds in eggs from hens after oral administration of mercury compounds. J. Agric. Food Chem. 17:1014-6.

Krista, L. M., C. W. Carlson, and O. E. Olson. 1961. Some effects of saline waters on chicks, laying hens, poults, and ducklings. Poult. Sci. 40:938-44.

Krista, L. M., C. W. Carlson, and O. E. Olson. 1962. Water for poultry. S. D. Farm Homes Res. 13 (No. 4):15-17.

Kroner, R. C., and J. F. Kopp. 1965. Trace elements in six water systems of the United States. J. Am. Water Works Assoc. 57:150-56.

Kunishisa, Y., T. Yaname, T. Tanaka, I. Fukuda, and T. Nishikava. 1966. The effect of dietary chromium on the performance of chicks. Jap. Poult. Sci. 3:10-14.

Larsen, C. and D. E. Bailey. 1913. Effect of alkali water on dairy cows, pp. 300-25. S. D. Agric. Exp. Sta. Bull. 147.

Lawlor, M. J., W. H. Smith, and W. M. Beeson. 1965. Iron requirement of the growing lamb. J. Anim. Sci. 24:742-47.

Lawrence, J. M. 1968. Comparative concentrations of selected macro and micro nutrients in water, suspended matter, hydrosol, plants and fish from reservoirs on Chattahoochee River, pp. 368-78. Proc. 4th Am. Water Res. Conf., Nov. 1. New York.

Leitch, I., and J. S. Thomson. 1944. The water economy of farm animals. Nutr. Abstr. Rev. 14(2):197-222.

LePore, Paul D., and R. F. Miller. 1965. Embryonic viability as influenced by excess molybdenum in chicken breeder diets. Proc. Soc. Exp. Biol. Med. 118:155-57.

Lewis, D. J. 1951a. The metabolism of nitrate and nitrite in sheep. I. The reduction of nitrate in the rumen of sheep. Biochem. J. 48:175-80.

Lewis, D. J. 1951b. The metabolism of nitrate and nitrite in sheep. II. Hydrogen donators in nitrate reduction by rumen microorganisms *in vitro*. Biochem. J. 49:149-53.

Lewis, G. C. 1959. Water quality study in the Boise Valley. Idaho Agric. Exp. Stn. Bull. No. 316. Moscow: University of Idaho.

Lichtenberg, J. J., J. W. Eichelberger, R. C. Dressman, and J. E. Longbottom. 1969. Pesticides in surface waters of the United States: A five year summary 1965-68. Pestic. Monit. J. 4:71-86.

Lieber, M., and W. F. Welsch. 1954. Contamination of ground water by cadmium. J. Am. Water Works. Assoc. 46:541-47.

Link, R. P., and R. R. Pensinger. 1966. Lead toxicosis in swine. Am. J. Vet. Res. 27:759-63.

Little, Arthur D., Inc. 1970. Water Quality Criteria Data Book, vol. 1. Organic chemical pollution of fresh water. Prepared for the Environmental Protection Agency, Water Quality Office, Washington, D.C.: U.S. Government Printing Office.

Livingstone, D. A. 1963. Chemical composition of rivers and lakes, Chapter G. Data for Geochemistry, 6th ed. USGS Prof. Pap. 440-G. Washington, D.C. 64 pp.

Lotspeich, F. B., V. L. Hauser, and O. R. Lehman. 1969. Quality of waters from playas on the Southern High Plains. Water Resour. Res. 5:48–58.

Lunin, Jesse. 1970. Agriculture's responsibility in establishing water quality. Talk presented at the 19th Annu. Meet. Agric. Res. Inst., Arlington, Va., Oct. 13–14.

Maag, D. D., and M. W. Glenn. 1967. Toxicity of selenium: Farm animals, pp. 127–40. In O. H. Muth, J. E. Oldfield, and P. H. Weswig, eds. Symposium: Selenium in Biomedicine. Westport, Conn.: AVI Publishing Co., Inc.

MacKenzie, R. D., R. U. Byerrum, C. F. Decker, C. A. Hoppert, and R. F. Langham. 1958. Chronic toxicity studies. II. Sexivalent and trivalent chromium administered in drinking water to rats. AMA Arch. Ind. Health 18:232–34.

Malan, A. I., P. J. DuToit, and J. W. Groenewald. 1935. Studies in mineral metabolism. 33. Iodine in nutrition of sheep. 2nd rep. Onderstepoort. J. Vet. Sci. 5:189–200.

Maletic, John T. 1973. Personal communication.

Martinez, A., and D. C. Church. 1970. Effect of various mineral elements on in vitro rumen cellulose digestion. J. Anim. Sci. 31:982–90.

Mason, K. E., and J. O. Young. 1967. Effectiveness of selenium and zinc in protecting against cadmium-induced injury of the rat testis, pp. 383–94. In O. H. Muth, J. E. Oldfield, and P. H. Weswig, eds. Symposium: Selenium in Biomedicine. Westport, Conn.: AVI Publishing Co., Inc.

Matrone, G., R. H. Hartman, and A. J. Clawson. 1959. Studies of Mn-Fe antagonism in the nutrition of rabbits and baby pigs. J. Nutr. 67:309–17.

Mayo, R. H., S. M. Hauge, H. E. Parker, F. N. Andrews, and C. W. Carrick. 1956. Copper tolerance of young chickens. Poult. Sci. 35:1156–57. (abstr.)

McKee, J. E., and H. W. Wolf. 1963. Water Quality Criteria, 2nd ed. Resources Agency of California State Water Quality Control Board, Publ. No. 3-A. Sacramento. 548 pp. (Reprinted in 1971.)

Medway, W., and M. R. Kare. 1958. The effect of excess salt when water intake is restricted. Rep. N.Y. State Vet. Coll. Cornell University, 1956–57. 23 pp.

Mehring, A. L., Jr., J. H. Brumbaugh, A. J. Sutherland, and H. W. Titus. 1960. The tolerance of growing chickens for dietary copper. Poult. Sci. 39:713–19.

Mertz, W. 1968. Problems in trace element research, pp. 163–69. In D. D. Hemphill, ed. Proc. 2nd Annu. Conf. Trace Subst. Environ. Health, July 16–18, 1968. Columbia: University of Missouri.

Miller, V. L., P. A. Klavano, A. C. Jerstad, and E. Csonka. 1961. Absorption distribution and excretion of ethylmercuric chloride. Toxicol. Appl. Pharmacol. 3:459–68.

Miller, W. J. 1971. Cadmium absorption, tissue and product distribution, toxicity effects and influence on metabolism of certain essential elements. Ga. Nutr. Conf., pp. 58–69.

Miller, W. J., B. Lampp, G. W. Powell, C. A. Salotti, and D. M. Blackman. 1967. Influence of a high level of dietary cadmium on cadmium content in milk, excretion and cow performance. J. Dairy Sci. 50:1404–8.

Mitchell, H. H. 1962. The water requirements for maintenance, Chapter 4. Comparative Nutrition of Man and Domestic Animals. New York: Academic Press. 701 pp.

Miyazaki, A. 1967. Studies on the effects of nitrate in feed upon the performance of ruminants. I. Effects of nitrate added to feed upon the gains and blood constituents of sheep. Jap. J. Zootech. Sci. 38:527 (Nutr. Abstr. 1968, 38(4):1412).

Moore, R. V., and G. W. Leddicotte. 1968. Trace substance interchange between sample and container—a significant problem in health-related research, pp. 243-506. In D. D. Hemphill, ed. Proc. 2nd Annu. Conf. Trace Subst. Environ. Health, July 16-18, 1968. Columbia: University of Missouri.

Morrison, F. B. 1936. Feeds and Feeding. Ithaca, N.Y.: The Morrison Publishing Co.

Morrison, F. B. 1959. Feeds and Feeding. Ithaca, N.Y.: The Morrison Publishing Co.

Mount, L. E., C. W. Holmes. W. H. Close, S. R. Morrison, and I. B. Start. 1971. A note on the consumption of water by the growing pig at several environmental temperatures and levels of feeding. Anim. Prod. 13:561-63.

Moxon, A. L., and M. Rhian. 1943. Selenium poisoning. Physiol. Rev. 23:305-37.

Moxon, A. L., and W. O. Wilson. 1944. Selenium-arsenic antagonism in poultry. Poult. Sci. 23:149-51.

Mugler, D. J., J. D. Mitchell, and A. W. Adams. 1970. Factors affecting turkey meat color. Poult. Sci. 49:1510-13.

National Academy of Sciences-National Academy of Engineering. Environmental Studies Board. 1972. Water Quality Criteria, Section 5, 594 pp.

National Research Council. 1966. Nutrient Requirements of Horses. Washington, D.C.: National Academy of Sciences.

National Research Council. 1968a. Nutrient Requirements of Sheep, 4th ed. Washington, D.C.: National Academy of Sciences. 64 pp.

National Research Council. 1968b. Nutrient Requirements of Swine, 6th ed. Washington, D.C.: National Academy of Sciences. 69 pp.

National Research Council. 1970a. Nutrient Requirements of Beef Cattle, 4th ed. Washington, D.C.: National Academy of Sciences. 55 pp.

National Research Council. 1971a. Nutrient Requirements of Dairy Cattle, 4th ed. Washington, D.C.: National Academy of Sciences. 54 pp.

National Research Council. 1971b. Nutrient Requirements of Poultry, 6th Rev. ed. Washington, D.C.: National Academy of Sciences. 54 pp.

National Research Council. 1971c. Selenium in nutrition. Washington, D.C.: National Academy of Sciences. 79 pp.

National Research Council. 1972a. Accumulation of Nitrate. Washington, D.C.: National Academy of Sciences. 106pp.

National Research Council. 1972b. Lead. Airborne lead in perspective. Washington, D.C.: National Academy of Sciences.

Nelson, L. W. 1966. Nitrite toxicosis and the gastric ulcer complex in swine. Part I: Nitrite toxicosis. Part II: Gastric ulcer complex. Diss. Abstr. 26:4140-41. East Lansing: Michigan State University.

Nelson, T. S., M. B. Gillis, and H. T. Peeler. 1962. Studies of the effect of vanadium on chick growth. Poult. Sci. 41:519-22.

Nesheim, M. C., R. M. Leach, Jr., T. R. Zeigler, and J. A. Serafin. 1964. Interrelationships between dietary levels of sodium, chlorine, and potassium. J. Nutr. 84:361-66.

Nicholson, H. P. 1970. The pesticide burden in water and its significance, Chapter 12. In T. L. Willrich and G. E. Smith, eds. Agricultural Practices and Water Quality. Ames: The Iowa State University Press.

Nicholson, J. W. G., J. K. Loosli, and R. G. Warner. 1960. Influence of mineral supplement on the growth of calves, digestibility of the rations and intraruminal environment. J. Anim. Sci. 19:1071-80.

Nugara, D., and H. M. Edwards, Jr. 1963. Influence of dietary Ca and P levels on the Mg requirement of the chick. J. Nutr. 80:181-84.

Ockerse, T. 1943. Endemic fluorosis in the Pretoria district. J. S. Afr. Med. 15:261; J. Am. Water Works Assoc. 35:242 (abstr.).

O'Dell, G. D., W. J. Miller, W. A. King, S. L. Moore, and D. M. Blackmon. 1970. Nickel toxicity in the young bovine. J. Nutr. 100:1447-53.

O'Donovan, P. B., R. A. Pickett, M. P. Plumlee, and W. M. Beeson. 1963. Iron toxicity in the young pig. J. Anim. Sci. 22:1075-80.

Officers of the Department of Agriculture and the Governmental Chemical Laboratories. 1950. Waters for agricultural purposes in Western Australia. J. Agric. West. Aust. 27(Series 2):156-60.

Olson, O. E., L. L. Sisson, and A. L. Moxon. 1940. Absorption of selenium and arsenic by plants from soils under natural conditions. Soil Sci. 50:115-18.

Ott, E. A., W. H. Smith, R. B. Harrington, and W. M. Beeson. 1966a. Zinc toxicity in ruminants. I. Effect of high levels of dietary zinc on gains, feed consumption and feed efficiency of lambs. J. Anim. Sci. 25:414-18.

Ott, E. A., W. H. Smith, R. B. Harrington, and W. M. Beeson. 1966b. Zinc toxicity in ruminants. II. Effect of high levels of dietary zinc on gains, feed consumption and feed efficiency of beef cattle. J. Anim. Sci. 25:419-23.

Ott, E. A., W. H. Smith, R. B. Harrington, M. Stob, H. E. Parker, and W. M. Beeson. 1966c. Zinc toxicity in ruminants. III. Physiological changes in tissues and alterations in rumen metabolism in lambs. J. Anim. Sci. 25:424-31.

Ott, E. A., W. H. Smith, R. B. Harrington, H. E. Parker, and W. M. Beeson. 1966d. Zinc toxicity in ruminants. IV. Physiological changes in tissues of beef cattle. J. Anim. Sci. 25:432-38.

Papworth, D. S. 1967. Organic compounds, Part Four. II. Pesticides, pp. 209-82. In E. G. C. Clarke and M. L. Clarke, eds. Garner's Veterinary Toxicology, 3rd ed. London: Bailliere, Tindall and Cassell, Ltd.

Pattison, E. S. 1970. Arsenic and water pollution hazard. Science 170:870-71.

Peirce, A. W. 1952. Studies on fluorosis of sheep. I. The toxicity of water-borne fluoride for sheep maintained in pens. Aust. J. Agric. Res. 3:326-40.

Peirce, A. W. 1957. Studies on salt tolerance of sheep. I. The tolerance of sheep for sodium chloride in the drinking water. Aust. J. Agric. Res. 8:711-22.

Peirce, A. W. 1959. Studies on salt tolerance of sheep. II. The tolerance of sheep for mixtures of sodium chloride and magnesium chloride in drinking water. Aust. J. Agric. Res. 10:725-35.

Peirce, A. W. 1960. Studies on salt tolerance of sheep. III. The tolerance of sheep for mixtures of sodium chloride and sodium sulfate in the drinking water. Aust. J. Agric. Res. 11:548-56.

Peirce, A. W. 1962. Studies on salt tolerance of sheep. IV. The tolerance of sheep for mixtures of sodium chloride and calcium chloride in the drinking water. Aust. J. Agric. Res. 13:479-86.

Peirce, A. W. 1963. Studies on salt tolerance of sheep. V. The tolerance of sheep for mixtures of sodium chloride, sodium carbonate, and sodium bicarbonate in the drinking water. Aust. J. Agric. Res. 14:815-23.

Peirce, A. W. 1966. Studies on salt tolerance of sheep. VI. The tolerance of wethers in pens for drinking waters of the types obtained from underground sources in Australia. Aust. J. Agric. Res. 17:209-18.

Peirce, A. W. 1968a. Studies on salt tolerance of sheep. VII. The tolerance of ewes and their lambs in pens for drinking waters of the types obtained from underground sources in Australia. Aust. J. Agric. Res. 19:577–87.

Peirce, A. W. 1968b. Studies on salt tolerance of sheep. VIII. The tolerance of grazing ewes and their lambs for drinking waters of the types obtained from underground sources in Australia. Aust. J. Agric. Res. 19:589–95.

Pensack, J. M., and R. C. Klussendorff. 1956. Newer aspects of zinc metabolism. Poult. Nutr. Conf., Atlantic City, N.J.

Peoples, S. A. 1964. Arsenic toxicity in cattle. Ann N. Y. Acad. Sci. 111:644–49.

Perry, H. M., Jr. 1968. Hypertension and trace minerals with particular emphasis on cadmium, pp. 101-25. In D. D. Hemphill, ed. Proc. 2nd Annu. Conf. Trace Subst. Environ. Health, July 16–18, 1968. Columbia: University of Missouri.

Pfander, W. H., G. B. Garner, W. C. Ellis, and M. E. Muhrer. 1957. The etiology of nitrate poisoning in sheep. Mo. Agric. Stn. Bull. 637. 12 pp.

Platnow, N., and H. K. Abbey. 1968. Toxicity of vanadium in calves. Vet. Res. 82:292.

Pletnikova, I. P. 1970. Biological effect and tolerance level of Se entering the organism in drinking water. Gig. Sanit. 35:14–19; Nutr. Abstr. 40(4):7650.

Plotnikov, K. I. 1960. Role of trace elements in the pathogenesis of gastrointestinal and pulmonary disorders of lambs on the Kulundinsk. Nutr. Abstr. Rev. 20:1138–39.

Pomelee, C. S. 1953. Toxicity of beryllium. Sewage Ind. Wastes 25:1424–28.

Powell, G. W., W. J. Miller, J. D. Morton, and C. M. Clifton. 1964. Influence of dietary cadmium level and supplemental zinc on cadmium toxicity in the bovine. J. Nutr. 84:205–13.

Radeleff, R. D. 1970. Veterinary Toxicology, 2nd ed. Philadelphia: Lea and Febiger.

Rainwater, F. H. 1962. Stream composition of the conterminous United States. USGS Hydrol. Invest. Atlas H-61. Washington, D.C.

Ramberg, C. F., Jr., J. M. Phang, G. P. Mayer, A. I. Norberg, and D. S. Kornfeld. 1970. Inhibition of calcium absorption and elevation of calcium removal rate from bone in fluoride-treated calves. J. Nutr. 100:981–89.

Ramsey, A. A. 1924. Waters suitable for livestock. Analyses and experiences in New South Wales. Agric. Gaz. N. S. Wales 35:339–42. (C.A. 18:3441).

Rankama, K., and T. A. Sahama. 1950. Geochemistry. Chicago: Chicago University Press. 927 pp.

Reid, G. K. 1961. Ecology of Inland Waters and Estuaries. New York: Reinhold Publishing Corp.

Roberson, R. H., and P. J. Schaible. 1960. The tolerance of growing chicks for high levels of different forms of zinc. Poult. Sci. 39:893–96.

Robinove, C. J., R. H. Langford, and J. W. Brookhart. 1958. Saline-water resources of North Dakota. USGS Water-Supply Paper 1428. Washington, D.C. 72 pp.

Robinson, J. R., and R. A. McCance. 1952. Water metabolism. Annu. Rev. Physiol. 14:115–42.

Romoser, G. L., W. A. Dudley, L. J. Machlin, and L. Loveless. 1961. Toxicity of vanadium and chromium for the growing chick. Poult. Sci. 40:1171–73.

Rosenfeld, G., and E. J. Wallace. 1953. Acute and chronic toxicity of germanium. Arch. Ind. Hyg. Occup. Med. 8:466–79.

Ross, D. B. 1964. Chronic copper poisoning in lambs. Vet. Rec. 76:875–76.

Sandstead, H. H., R. F. Burk, G. H. Booth, Jr., and W. J. Darby. 1970. Current concepts of trace minerals. Clinical considerations. Med. Clin. N. Am. 54:1509–31.

Saville, P. D. 1967. Water fluoridation: Effect on bone fragility and skeletal calcium content in the rat. J. Nutr. 91:353–57.

Schroeder, H. A., W. H. Vinton, Jr., and J. J. Balassa. 1963a. Effect of chromium, cadmium and other trace metals on the growth and survival of mice. J. Nutr. 80:39–47.

Schroeder, H. A., W. H. Vinton, Jr., and J. J. Balassa. 1963b. Effects of chromium, cadmium and lead on the growth and survival of rats. J. Nutr. 80:48–54.

Schroeder, H. A., J. J. Balassa, and W. H. Vinton, Jr. 1965. Chromium, cadmium, and lead in rats: Effects of life span, tumors and tissue levels. J. Nutr. 86:51–66.

Schroeder, H. A., and J. J. Balassa. 1967. Arsenic, germanium, tin and vanadium in mice: Effects of growth, survival and tissue levels. J. Nutr. 92:245–52.

Schroeder, H. A., M. Mitchener, J. J. Balassa, M. Kanisawa, and A. P. Nason. 1968a. Zirconium, niobium, antimony and fluorine in mice: Effects on growth, survival and tissue levels. J. Nutr. 95:95–101.

Schroeder, H. A., M. Kanisawa, D. V. Frost, and M. Mitchener. 1968b. Germanium, tin and arsenic in rats: Effects on growth, survival, pathological lesions and life span. J. Nutr. 96:37–45.

Schroeder, H. A., and M. Mitchener. 1971a. Selenium and tellurium in rats: Effects on growth, survival and tumors. J. Nutr. 101:1531–40.

Schroeder, H. A., and M. Mitchener. 1971b. Toxic effects of trace elements on the reproduction of mice and rats. Arch. Environ. Health 23:102–6.

Schwartz, C. C., J. G. Nagy, and C. L. Streeter. 1973. Pesticide effect on rumen microbial function. J. Anim. Sci. 37:821–26.

Scrivner, L. H. 1946. Experimental edema and ascites in poults. J. Am. Vet. Med. Assoc. 108:27–32.

Seerley, R. W., R. J. Emerick, L. B. Embry, and O. E. Olson. 1965. Effect of nitrate or nitrite administered continously in drinking water for swine and sheep. J. Anim. Sci. 24:1014–19.

Sell, J. L., and W. K. Roberts. 1963. Effects of dietary nitrate on the chick: Growth, liver vitamin A stores and thyroid weight. J. Nutr. 79:171–78.

Selye, H. 1943. Production of nephrosclerosis in the fowl by sodium chloride. J. Am. Vet. Med. Assoc. 103:140–43.

Shand, A., and G. Lewis. 1957. Chronic copper poisoning in young calves. Vet. Rec. 69:618–20.

Shilo, M. 1967. Formation and mode of action of algal toxins. Bacteriol. Rev. 31:180–93.

Shirley, R. L. 1970. Nutrients in water available to economic animals, pp. 23–25. Proc. Nutr. Counc. Annu. Meet. Am. Feed Manuf. Assoc., Chicago, Ill., May.

Shirely, R. L., G. K. Davis, and J. R. Neller. 1951. Distribution of P^{32} in the tissues of a steer fed grass from land that received labelled fertilizer. J. Anim. Sci. 10:335–36.

Shirley, R. L., W. K. Robertson, J. T. McCall, J. R. Neller, and G. K. Davis. 1957. Distribution of Ca^{45} in tissue of a steer fed grass from land that received labelled fertilizer. Q. J. Fla. Acad. Sci. 20:133–38.

Shupe, J. L., M. L. Miner, and D. A. Greenwood. 1964. Clinical and pathological aspects of fluorine toxicosis in cattle. Ann. N. Y. Acad. Sci. 111:618–37.

Simon, J., J. M. Sund, F. D. Douglas, M. J. Wright and T. Kowalczyk. 1959. The effect of nitrate or nitrite when placed in the rumens of pregnant dairy cattle. J. Am. Vet. Med. Assoc. 135:311–14.

Sinclair, K. B., and D. I. H. Jones. 1967. Nitrite toxicity in sheep. Res. Vet. Sci. 8:65-70.

Sleight, S. D., and O. A. Atallah. 1968. Reproduction in the guinea pig as affected by chronic administration of potassium nitrate and potassium nitrite. Toxicol. Appl. Pharmacol. 12:179-85.

Sloan, C. E. 1970. Biotic and hydrologic variables in prairie potholes in North Dakota. J. Range Manage. 23(4):260-63.

Smith, G. E. 1965. Nitrate problems in water as related to soils, plants and water. Mo. Agric. Exp. Stn. Spec. Rep. 55:42-52.

Smith, O. M. 1944. The detection of poisons in public water supplies. Water Works Eng. 97:1293-1312.

Sokolowski, J. H. 1966. Nitrate poisoning: A chemical and biological evaluation of potassium nitrate utilization by the growing, fattening lamb. Diss. Abstr. 26:6946-47.

Spafford, W. J. 1941. South Australian natural waters for farm livestock. J. Dep. Agric. S. Aust. 44:619-28.

Standish, J. F., C. B. Ammerman, C. F. Simpson, F. C. Neal, and A. Z. Palmer. 1969. Influence of graded levels of dietary Fe as ferrous sulfate on performance and tissue mineral composition of steers. J. Anim. Sci. 29:496-503.

Stickel, W. H., L. F. Stickel, and F. B. Coon. 1970. DDE and DDD residues correlated with mortality of experimental animals, pp. 287-94. In: W. E. Deichmann, L. L. Radomski, and R. A. Panalver, eds. Pesticides Symposia. Inter-American Conferences on Toxicology and Occupational Medicine. Miami, Fla.: Halos and Assoc., Inc.

Stokinger, H. E. 1969. The spectre of today's environmental pollution USA brand: New perspectives from an old scout. Cummings Memorial Lecture. Am. Ind. Hyg. Assoc. J. 30:195-217.

Stokinger, H. E., and R. L. Woodward. 1958. Toxicologic methods for establishing drinking water standards. J. Am. Water Works Assoc. 50:515-29.

Storer, N. L., and T. S. Nelson. 1968. The effect of various aluminum compounds on chick performance. Poult. Sci. 47:244-47.

Stormoken, H. 1953. Methemoglobinemia in domestic animals. Proc. Fifteenth Int. Vet. Congr.

Sturkie, P. D. 1956. The effects of excess zinc on water consumption in chickens. Poult. Sci. 35:1123-24.

Sunde, M. L. 1964. The use of the turkey poult as a test animal for nitrite toxicity. Poult. Sci. 43:1368. (abstr.)

Sunde, M. L. 1967. Water is important. Feedstuffs 39:22.

Sunde, M. L. 1971. Personal communication.

Supplee, W. C. 1961. Production of zinc deficiency in turkey poults by dietary cadmium. Poult. Sci. 40:827-28.

Swensson, A., K. D. Lundgren, and O. Lindstrom. 1959. Retention of various mercury compounds after subacute administration. AMA Arch. Ind. Health 20:467-72.

Sykes, J. F. 1955. Animals and fowl and water, pp. 14-18. The Yearbook of Agriculture. USDA. Washington, D.C.: Government Printing Office.

Systems for Technical Data (STORET), Water Programs Office, Environmental Protection Agency, Washington, D.C. 1971.

Tanner, J. T., M. H. Friedman, D. N. Lincoln, L. A. Ford, and M. Jaffee. 1972.

Mercury content of common foods determined by neutron activation analysis. Science 177:1102–03.

Taucins, E., A. Svilane, A. Valdmanis, A. Buike, R. Zarina, and E. Fedorova. 1969. Fiziol. Akt. Komponenty Pitan. Zhivotn. 199–212.

Thomas, J. W., and S. Moss. 1951. The effect of orally administered Mo on growth, spermatogenesis and testes histology of young dairy bulls. J. Dairy Sci. 34:929–34.

Timmons, F. L., P. A. Frank, and R. J. Demint. 1970. Herbicide residues in agricultural water from control of aquatic and bank weeds, Chapter 13. In T. L. Willrich and G. E. Smith, eds. Agricultural Practices and Water Quality. Ames: The Iowa State University Press.

Toetz, D. W. 1967. Ecological factors affecting turbidity and productivity in prairie ponds in the Southern Great Plains. Water Resources Research Institute, Oklahoma State University, Stillwater.

Tollett, J. T., D. E. Becker, A. H. Jensen, and S. W. Terrill. 1960. Effect of dietary nitrate on growth and reproductive performance of swine. J. Anim. Sci. 19:1297. (abstr.)

Turk, J. L., Jr., and F. H. Kratzer. 1960. The effects of cobalt in the diet of the chick. Poult. Sci. 39:1302. (abstr.)

Uchida, M., K. Hirakawa, and T. Inoue. 1961. Biochemical studies on Minamata disease. III. Relations between the causal agent of the disease and the mercury compound in the shellfish with reference to their chemical behaviors. Kumamoto Med. J. 14:171–79. (C.A. 57:10378e).

Ulfvarson, U. 1969. Absorption and distribution of mercury in rats fed organs from rats injected with various mercury compounds. Toxicol. Appl. Pharmacol. 15:525–31.

Underwood, E. J. 1971. Trace Elements in Human and Animal Nutrition, 3rd ed. New York: Academic Press.

United States Department of Health, Education, and Welfare. 1962a. Drinking Water Standards, Rev. ed. PHS Publ. No. 956. Washington, D.C.: U.S. Government Printing Office.

United States Department of Health, Education, and Welfare. 1962b. Drinking Water Standards. Part 72, Interstate Quar. Fed. Regist. 2152. (March 6)

United States Federal Radiation Council. 1961. Background material for the development of radiation protection standards, staff report. Washington, D. C.: U.S. Government Printing Office. 19 pp.

United States Geological Survey Water-Supply Paper 1643. 1959a. Quality of Surface Waters of the United States, 1959, Parts 5 and 6. Hudson Bay and Upper Mississippi River Basins, and Missouri River Basin. S. K. Love, ed. Washington, D.C. 247 pp.

United States Geological Survey Water-Supply Paper 1645. 1959b. Quality of Surface Waters of the United States, 1959, Parts 9–14. Colorado River Basin to Pacific Slope Basins in Oregon and Lower Columbia River Basin. Washington, D.C. 523 pp.

Unites States Geological Survey Water-Supply Paper 1743. 1960. Quality of Surface Waters of the United States, 1960, Parts 5 and 6. Hudson Bay and Upper Mississippi River Basins and Missouri River Basin. Washington, D.C. 278 pp.

United States Geological Survey Water-Supply Paper 1884. 1961a. Quality of Surface Waters of the United States, 1961, Parts 7 and 8. Lower Mississippi River Basins and Western Gulf of Mexico Basins. Washington, D.C. 590 pp.

United States Geological Survey Water-Supply Paper 1885. 1961b. Quality of Surface Waters of the United States, 1961, Parts 9–14. Colorado River Basin to Pacific Slope Basins in Oregon and Lower Columbia River Basin. Washington, D.C. 677 pp.

United States Public Health Service. 1964–65. Air Quality Data. United States Department of Health, Education, and Welfare. Washington, D.C. 125 pp.

Utley, P. R., N. W. Bradley, and J. A. Boling. 1970. Effect of water restriction on nitrogen metabolism in bovines fed two levels of nitrogen. J. Nutr. 100:551–56.

Vohra, P., and F. H. Kratzer. 1968. Zinc, copper and manganese toxicities in turkey poults and their alleviation by EDTA. Poult. Sci. 47:699–704.

Wahlstrom, R. C., and O. E. Olson. 1959a. The effect of selenium on reproduction in swine. J. Anim. Sci. 18:141–45.

Wahlstrom, R. C., and O. E. Olson. 1959b. The relation of pre-natal and pre-weaning treatment to the effect of arsanilic acid on selenium poisoning in weanling pigs. J. Anim. Sci. 18:578–82.

Waibel, P. E., D. C. Snetsinger, R. A. Ball, and J. H. Sautter. 1964. Variation in tolerance of turkeys to dietary copper. Poult. Sci. 43:504–6.

Waldron, A. C., H. E. Kaeser, D. L. Coleman, J. R. Staubus, and H. D. Niemczyk. 1968. Heptachlor and heptachlor epoxide residues on fall-treated alfalfa and in milk and cow tissues. J. Agric. Food Chem. 16:627–31.

Warren, H. V., R. E. Delavault, and C. H. Cross. 1969. Base metal pollution in soils, pp. 9–19. In D. D. Hamphill, ed. Proc. 3rd Annu. Conf. Trace Subst. Environ. Health, July 16–18, 1968. Columbia: University of Missouri.

Water Quality Criteria. 1968. Section IV, Agricultural uses, pp. 112–177. Rep. Nat. Tech. Advis. Comm. Secr. USDI. Washington, D.C.: U.S. Government Printing Office.

Way, J. M. 1969. Toxicity and hazards of auxin herbicides. Residue Rev. 26:37–62.

Weber, C. W., and B. L. Reid. 1968. Nickel toxicity in growing chicks. J. Nutr. 95:612–16.

Weber, C. W., and B. L. Reid. 1969. Nickel toxicity in young growing mice. J. Anim. Sci. 28:620–23.

Weeth, H. J. 1962. Effect of drinking water containing added manganese on cattle. J. Anim. Sci. 21:656. (abstr.)

Weeth, H. J., and D. L. Caps. 1971. Tolerance of cattle for sulfate water. J. Anim. Sci. 33:211–12. (abstr.)

Weeth, H. J., L. H. Haverland, and D. W. Cassard. 1960. Consumption of sodium chloride water by heifers. J. Anim. Sci. 19:845–51.

Weeth, H. J., and J. E. Hunter. 1971. Drinking of sulfate-water by cattle. J. Anim. Sci. 32:277–81.

Weeth, H. J., and A. L. Lesperance. 1965. Renal function of cattle under various water and salt loads. J. Anim. Sci. 24:441–47.

Weeth, H. J., A. L. Lesperance, and V. R. Bohman. 1968. Intermittent saline watering of growing beef heifers. J. Anim. Sci. 27:739–44.

Weichenthal, B. A., L. B. Emery, R. J. Emerick, and F. W. Whetzal. 1963. Influence of sodium nitrate, vitamin A and protein level on feedlot performance and vitamin A status of fattening cattle. J. Anim. Sci. 22:979–84.

Wells, D. M., R. G. Rekers, and E. W. Huddleston. 1970. Potential pollution of the Ogallala by recharging playa lake water. Proj. Completion Rep. WRC-70-4. Water Resources Center, Texas Tech University, Lubbock.

REFERENCES 93

White, D. E., J. D. Hem, and G. A. Waring. 1963. Chemical composition of subsurface waters, Chapter F. Data of Geochemistry. USGS Prof. Pap. 440-F. Washington, D.C.: U.S. Government Printing Office. 67 pp.

Willrich, T. L., and G. E. Smith, eds. 1970. Agricultural Practices and Water Quality. Ames: The Iowa State University Press. 415 pp.

Winchester, C. F., and M. J. Morris. 1956. Water intake rates in cattle. J. Anim. Sci. 15:722-40.

Winks, W. R., A. K. Sutherland, and R. M. Salisbury. 1950. Nitrite poisoning of pigs. Queensland J. Agric. Sci. 7:1-14.

Winnek, P. S., and A. H. Smith. 1937. Studies in the role of Br in nutrition. J. Biol. Chem. 121:345-52.

Winter, A. J., and J. F. Hokanson. 1964. Effects of long-term feeding of nitrate, nitrite, or hydroxylamine on pregnant dairy heifers. Am. J. Vet Res. 25:353-61.

Young, D. R., N. S. Schafer, and R. Price. 1960. Effect of nutrient supplements during work on performance capacity of dogs. J. Appl. Physiol. 15:1022-26.

*2284-19
1977
5-30
C